SOME IMPORTANT FACTS ABOUT THIS REVISION GUID

- IT IS MEANT TO BE USED THROUGHOUT YOUR COURSE AS WELL AS FOR PRE-EXAM REVISION.
- IT IS WRITTEN SPECIFICALLY FOR YOUR SYLLABUS.
- IT CONTAINS ALL THE INFORMATION YOU NEED TO KNOW ...
- ... PRESENTED IN AN INFORMAL, USER FRIENDLY STYLE.
- TOPICS ARE PRESENTED AS SINGLE OR DOUBLE PAGE SPREADS ...
- ... WHICH ARE THEN DIVIDED INTO EASY TO LEARN SUB SECTIONS.
- NOTES ARE WRITTEN IN SHORT, SNAPPY PHRASES ...
- ... WITH KEY FEATURES ...

... HIGHLIGHTED BY GREY BOXES ...

... WRITTEN IN CAPITALS ...

... OR EMPHASISED BY 'BULLET POINTS'.

- AT THE END OF EACH PAGE THERE IS A KEY POINTS SECTION ...
- ... WITH A FULL SET OF SUMMARY QUESTIONS AT THE END OF EACH THEME.

THOSE SECTIONS OF THE GUIDE WHICH ARE OUTLINED IN RED ARE THE AREAS OF THE SYLLABUS WHICH WILL NOT BE TESTED ON THE FOUNDATION TIER WRITTEN EXAMINATION ...
... AND SHOULD ONLY BE LEARNED BY PUPILS ENTERED FOR THE HIGHER TIER EXAMINATION.

SOME IMPORTANT FACTS ABOUT YOUR EXAMINATION

- You will have THREE PAPERS, lasting 1 HOUR 30 MINUTES EACH.
- All papers will consist of COMPULSORY QUESTIONS of different lengths, providing opportunity for extended prose writing.
- Each paper carries 100 marks.

PAPER 1	LIFE AND LIVING PROCESSES.
PAPER 2	MATERIALS AND THEIR PROPERTIES.
PAPER 3	PHYSICAL PROCESSES.

HOW TO USE THIS REVISION GUIDE

- Don't just read! LEARN ACTIVELY!
- Constantly test yourself ... WITHOUT LOOKING AT THE BOOK.
- When you have revised a small sub-section or a diagram, PLACE A BOLD TICK AGAINST IT. Similarly, fill in the "Progress and Revision" chart on Page 52.
- Jot down anything which will help YOU to remember - no matter how trivial it may seem.
- **DON'T BE TEMPTED TO COLOUR IN DIAGRAMS. COLOUR CAUSES CONFUSION - YOUR EXAM WILL BE IN BLACK AND WHITE !**
- These notes are highly refined. Everything you need is here, in a highly organised but user friendly format. Many questions depend only on STRAIGHTFORWARD RECALL OF FACTS, so make sure you LEARN THEM.
- THIS IS YOUR BOOK! Use it throughout your course in the ways suggested and your revision will be both organised and successful.

A WORD ABOUT THE LAYOUT OF THE PHYSICS GUIDE

We have reduced the number of themes from 6 to 5 by including 'the Earth and beyond' in the 'Forces and Motion' theme. Nevertheless, the contents page links each page to its syllabus reference number and the page titles are also easily linked to the syllabus.

CONTENTS

* Numbers in brackets refer to syllabus reference No.

NOTES

FORCES AND MOTION

SPEED AND VELOCITY — FM 1

DISTANCE TRAVELLED IN METRES, ... EVERY SECOND

The relationship:

$$\text{SPEED (m/s)} = \frac{\text{DISTANCE (m)}}{\text{TIME TAKEN (s)}}$$

FORMULA TRIANGLE: d over $s \times t$

Usually measured in m/s, Km/h or miles/h

N.B. **VELOCITY OF AN OBJECT IS ITS SPEED IN A PARTICULAR DIRECTION**

YOU MUST KNOW THIS FORMULA - SO LEARN IT

CONSTANT or UNIFORM SPEED

- e.g. cyclist travelling at 8m/s

1s 1s 1s

8m 8m 8m

- Same DISTANCE ... every SECOND ... CONSTANT SPEED.

EXAMPLE: "Calculate the speed of a cyclist who travels 2400m in 5mins."

$$\text{SPEED} = \frac{\text{DISTANCE TRAVELLED}}{\text{TIME TAKEN}} = \frac{2400\text{(m)}}{300\text{(secs remember!)}} = 8\text{m/s}$$

DISTANCE - TIME GRAPHS

1. STATIONARY

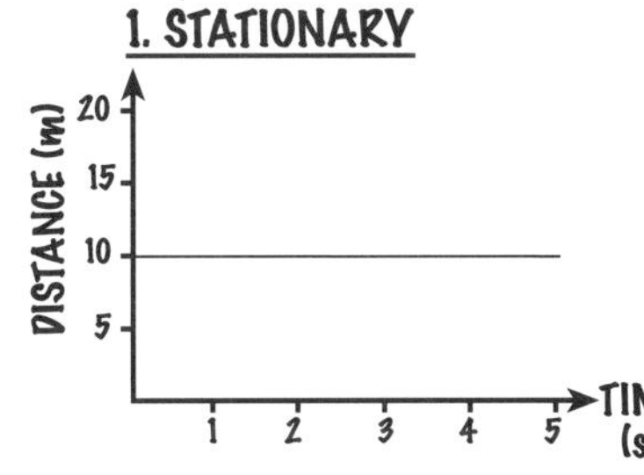

2. MOVING AT CONSTANT SPEED

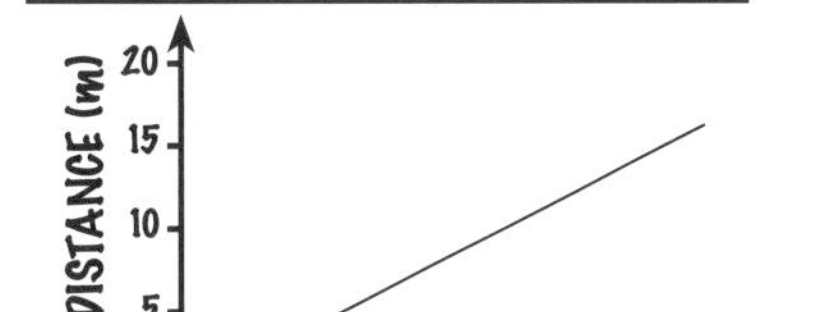

3. MOVING AT GREATER CONSTANT SPEED

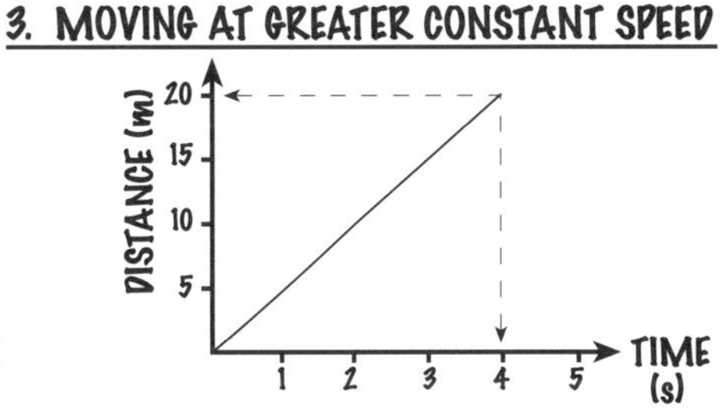

- The SLOPE of a DISTANCE-TIME GRAPH is a measure of the SPEED OF THE OBJECT ...
- ... and the STEEPER the SLOPE, the GREATER the SPEED.

EXAMPLE

For No. 3, $\text{GRADIENT} = \text{SPEED} = \frac{\text{DISTANCE TRAVELLED}}{\text{TIME TAKEN}} = \frac{20\text{m}}{4\text{s}} = 5\text{m/s}$ (see graph above).

DISTANCE - TIME GRAPHS FOR CHANGING VELOCITY AND ACCELERATION

1. CHANGING VELOCITY

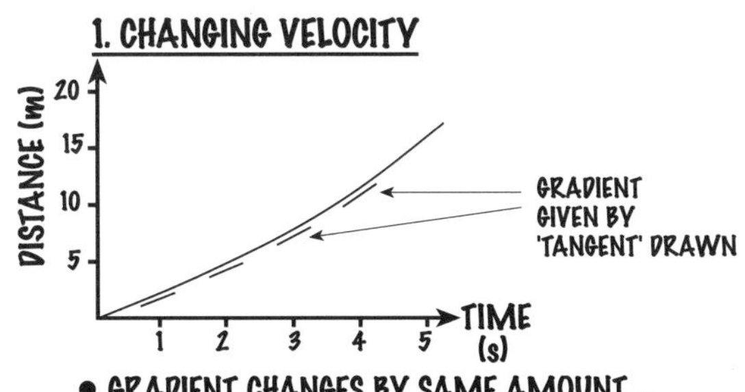

- GRADIENT CHANGES BY SAME AMOUNT ...
- ... EVERY SECOND and so
- VELOCITY CHANGES BY SAME AMOUNT ...
- ... EVERY SECOND
- OBJECT MOVING WITH CONSTANT CHANGING VELOCITY.

2. CHANGING ACCELERATION

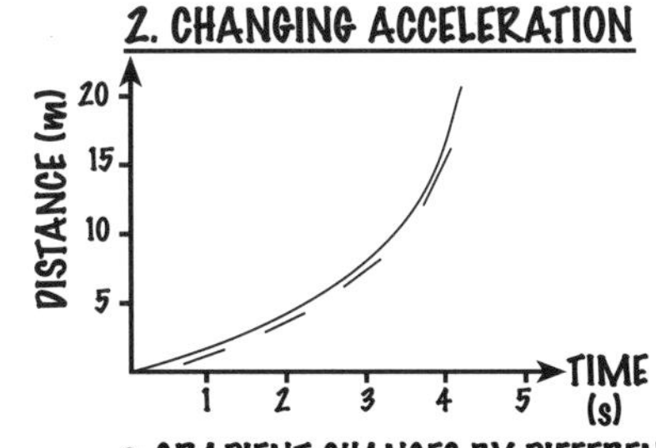

- GRADIENT CHANGES BY DIFFERENT AMOUNTS ...
- ... EVERY SECOND and so
- VELOCITY CHANGES BY DIFFERENT AMOUNTS ...
- ... EVERY SECOND
- OBJECT MOVING WITH CHANGING ACCELERATION.

KEY POINTS:

- Speed = Distance ÷ Time taken
- Measured in m/s, Km/h or miles/h.
- Velocity is speed in a particular direction.
- Distance-time graphs can be used to represent the motion of an object.

FORCE AND ACCELERATION I

- Forces are PUSHES or PULLS e.g. FRICTION, WEIGHT, AIR RESISTANCE.
- They are measured in NEWTONS (N) and may be different in SIZE and DIRECTION.
- Whenever TWO BODIES INTERACT i.e. are in contact, they exert EQUAL and OPPOSITE FORCES on each other ...

... e.g. when you stand on any surface you exert a DOWNWARD FORCE called WEIGHT on that surface ...
... while the surface exerts an EQUAL and UPWARD FORCE ON YOU!!!

HOW FORCES AFFECT MOVEMENT - a very simple consideration

1. BALANCED FORCES - EQUAL AND OPPOSITE

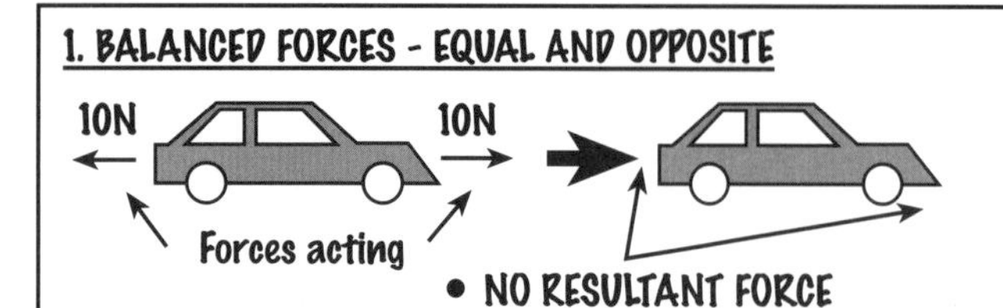

EFFECT ON MOVEMENT -

- If already STATIONARY ... it remains STATIONARY.
- If already MOVING ...
  ... it continues at the SAME SPEED, ...
  ... in the SAME DIRECTION.
  ... i.e. its VELOCITY is UNCHANGED

2. UNBALANCED FORCES - UNEQUAL AND OPPOSITE

- RESULTANT FORCE ...
- SIZE IS DIFFERENCE BETWEEN FORCES.
- DIRECTION IS SAME AS LARGEST INITIAL FORCE.

EFFECT ON MOVEMENT -

- If already STATIONARY ... it begins to move
  ... its VELOCITY is CHANGED.
- If already MOVING ...
  ... its VELOCITY INCREASES (ACCELERATES)
  ... or its VELOCITY DECREASES (DECELERATES)

FRICTION

FRICTION is a FORCE that OPPOSES THE DIRECTION OF MOVEMENT of an OBJECT ...
... and it acts when an object MOVES THROUGH A MEDIUM e.g. air, water ...
... or when SURFACES SLIDE past each other.

VEHICLES EXPERIENCE FRICTION BETWEEN THEIR TYRES AND ROAD SURFACE.

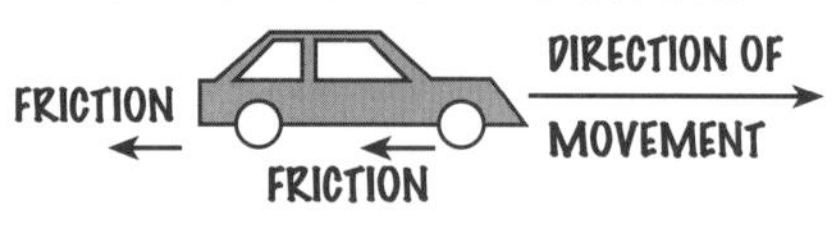

- For movement to occur from rest ...
- ... FORCE OF ENGINE MUST BE GREATER THAN FORCE OF FRICTION.

FRICTION (AIR RESISTANCE)

DIRECTION OF MOVEMENT

A PARACHUTE EXPERIENCES FRICTION BETWEEN THE PARACHUTE AND THE AIR.

This friction force is called ...

- ... AIR RESISTANCE or DRAG ...
- ... which results in the SPEED ...
- ... of the parachutist being REDUCED ...
- ... to a SAFE and CONSTANT LEVEL.

N.B. THE FASTER THE OBJECT MOVES, THE GREATER THE FORCE OF FRICTION.

STOPPING DISTANCES

There are many factors which affect the STOPPING DISTANCE of a vehicle:

STOPPING DISTANCE INCREASED

1. FASTER THE VEHICLE IS TRAVELLING (fairly obvious)
2. POOR REACTION TIME OF DRIVER (due to TIREDNESS, ALCOHOL, DRUGS)
3. LACK OF FORCE APPLIED BY BRAKING SYSTEM (due to worn brake shoes, brake pedal not pushed enough)
4. LACK OF FRICTION BETWEEN WHEELS AND SURFACE (due to BALD TYRES, WET or ICY SURFACE)

KEY POINTS:

- A force is a push or a pull and is measured in Newtons (N).
- The movement of an object depends on whether the forces acting on it are balanced or unbalanced.

ACCELERATION - rate of change of velocity

Very simply ... • CHANGE OF VELOCITY IN METRES/SEC ... EVERY SECOND.
• IF VELOCITY GOES UP = ACCELERATION. IF VELOCITY GOES DOWN = DECELERATION.

CONSTANT ACCELERATION • e.g. cyclist accelerating at $2m/s^2$ (initially at rest)

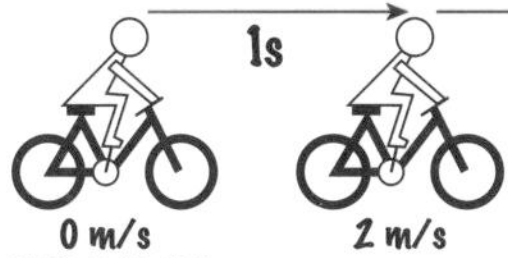

1s 1s

4 m/s

• VELOCITY goes up ... by same amount ... every SECOND ... CONSTANT ACC^{N} ... distance NOT constant

The Relationship: ACCELERATION (m/s^2) = $\frac{\text{CHANGE IN VELOCITY (m/s)}}{\text{TIME TAKEN (s)}}$

Only one unit, m/s^2

YOU MUST KNOW THIS FORMULA - SO LEARN IT

$\frac{(v-u)}{a \times t}$

• ... where (v-u) is CHANGE IN VELOCITY.

EXAMPLE: A cyclist accelerates uniformly from rest and reaches a velocity of 20m/s after 5s, before decelerating uniformly and coming to rest in a further 10s. Calculate a) his ACCELERATION, and b) his DECELERATION.

(a) ACC$^{N.}$ = $\frac{\text{CHANGE IN VELOCITY}}{\text{TIME TAKEN}}$ = $\frac{20 - 0}{5}$ = $4m/s^2$

(b) DECEL$^{N.}$ = $\frac{\text{CHANGE IN VELOCITY}}{\text{TIME TAKEN}}$ = $\frac{20 - 0}{10}$ = $2m/s^2$ i.e. a DECELERATION

make sure you state this!

FORCE, MASS AND ACCELERATION

As we have seen the MOVEMENT of an OBJECT depends on whether the FORCES acting on it are BALANCED or UNBALANCED. Here we will deal with the effect in more detail.

The ACCELERATION of an object depends on ...

- The RESULTANT FORCE acting - the bigger the force, the bigger the acceleration.
- the MASS of the object - the bigger the mass, the smaller the acceleration.

The relationship:- FORCE (NEWTONS) = MASS (Kg) x ACCELERATION (m/s^2)

YOU MUST KNOW THIS FORMULA - SO LEARN IT

$\frac{F}{m \times a}$

EXAMPLE: A toy bike of mass 500g is pulled along a floor with a constant speed by a force of 5N. The pull is increased and the bike accelerates at $2m/s^2$. Calculate,
a) the force needed to achieve this acceleration b) the total pull force exerted on the bike.

A diagram is always a good start!

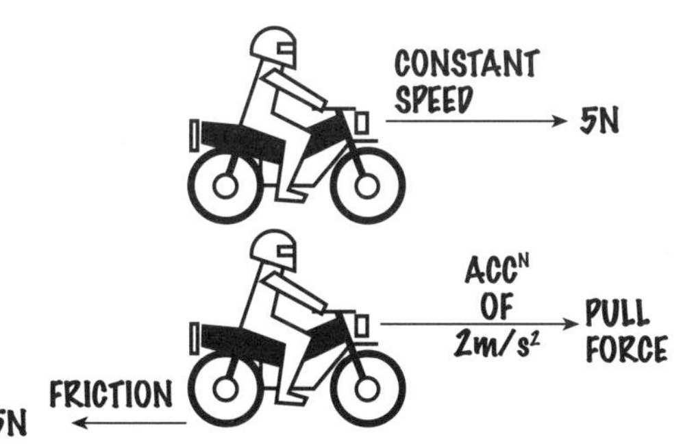

- As the bike is moving at CONSTANT SPEED ...
- ... the forces acting must be BALANCED.
- Therefore the 5N pull must be opposed by ...
- ... an EQUAL FORCE i.e. FRICTION/AIR RESISTANCE.

- As the bike is now ACCELERATING ...
- ... the pull force must be greater than the friction etc.
- i.e. the forces are now UNBALANCED ...
- ... and a RESULTANT FORCE now acts.

(a) Using our equation: F = m x a
F = 0.5kg x $2m/s^2$
F = 1 NEWTON.

MUST BE IN Kg

(b) TOTAL "pull force" = Force to provide acceleration + force needed to overcome friction.
= 1N + 5N i.e. 6 NEWTONS

KEY POINTS:

• Acceleration (m/s^2) = Change in velocity (m/s) ÷ Time taken (s). • Acceleration of an object depends on the resultant force acting and mass of object. • Force (N) = Mass (Kg) x Acceleration (m/s^2)

MASS, WEIGHT AND GRAVITY

Basically ...
- The MASS in KILOGRAMS (Kg) of an OBJECT is the AMOUNT OF MATTER it contains ...
- ... and the WEIGHT in NEWTONS (N) of an OBJECT is the DOWNWARD FORCE ...
- ... on that object due to GRAVITY.

Where ...
- GRAVITY is the FORCE of ATTRACTION that exists between ANY TWO OBJECTS ...
- ... in this case the OBJECT and the EARTH.

While ...
- Near the surface of the earth EVERY 1KG OF MATTER has a WEIGHT of 10N.

ACCELERATION DUE TO GRAVITY

The accn of a FREELY FALLING OBJECT on earth depends on whether the object is falling through ...
- A VACUUM.
- AIR or ANOTHER FLUID (e.g. water).

SINCE ALL FALLING OBJECTS EXPERIENCE A DOWNWARDS FORCE (GRAVITY) ...
- ... THEN WHILE FALLING THEY MUST ALSO BE ACCELERATING however ...

... in AIR and OTHER FLUIDS

- OBJECT DOES NOT experience ...
- ... an OPPOSING UPWARD FORCE ...
- ... i.e. FRICTION and so ...
- ... ALL objects fall with a ...
- ... CONSTANT ACC^n in a VACUUM.
- ... Its value is 10m/s^2

- OBJECT DOES experience ...
- ... an OPPOSING UPWARD FORCE ...
- ... due to FRICTION between the OBJECT ...
- .... and the AIR or FLUID it falls through.
- Therefore it will ONLY ACCELERATE UP TO A POINT! (see below).

TERMINAL VELOCITY - a freefalling experience

This is the CONSTANT VELOCITY reached by FALLING OBJECTS in AIR and OTHER FLUIDS (not in a VACUUM!) and so ...

The next time you jump out of a plane remember ... the FORCES acting on you are ...
- WEIGHT, W (↓) and • FRICTION DUE TO AIR called AIR RESISTANCE or DRAG, R (↑).

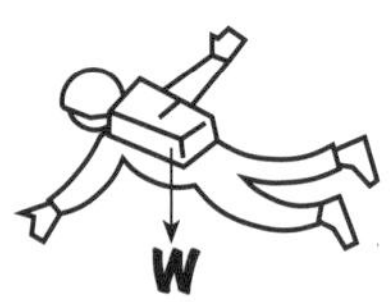

R
W

R
W

- INITIALLY ...
- ... only force acting is W ...
- ... air resistance R, experienced ...
- ... as soon as you fall!

- As you fall, you accelerate ...
- ... because W > R
- However, as you accelerate, ...
- ... R increases, but W doesn't!

- EVENTUALLY ...
- R increases until R = W ...
- ... meaning no resultant force so, ...
- ... no acceleration. TERMINAL VELOCITY REACHED

Factors which affect terminal velocity

1. AIR RESISTANCE
- the GREATER the AIR RESISTANCE ...
- ... the MORE RESISTANCE there is AGAINST MOVEMENT.
- It then takes LESS TIME until R=W which means ...
- ... object ACCELERATING FOR A SHORTER TIME and so ...
- ... TERMINAL VELOCITY of object is REDUCED.

2. SHAPE OF OBJECT
- STREAMLINED OBJECTS offer ...
- ... LESS RESISTANCE AGAINST MOVEMENT.
- It then takes MORE TIME until R=W which means ...
- ... object ACCELERATING FOR A LONGER TIME and so ...
- ... TERMINAL VELOCITY of object is INCREASED.

KEY POINTS:

- Gravity gives an object weight. • A mass of 1kg has a weight of 10N near the surface of the earth.
- An object falling through a medium (e.g. air) will reach terminal velocity.

FORCE ON SOLIDS

- FORCES can be used to TURN and STRETCH OBJECTS and to APPLY PRESSURE TO THEM, also...
- ... FORCES can be TRANSFERRED through LIQUIDS by APPLYING PRESSURE to them ...
- ... while APPLYING PRESSURE to GASES FORCES THEM into a SMALLER SPACE OR VOLUME.

TURNING EFFECT OF A FORCE - MOMENTS

Depends on
- SIZE OF APPLIED FORCE - greater force, greater turning effect.
- PERPENDICULAR DISTANCE FROM PIVOT - greater distance, greater turning effect.

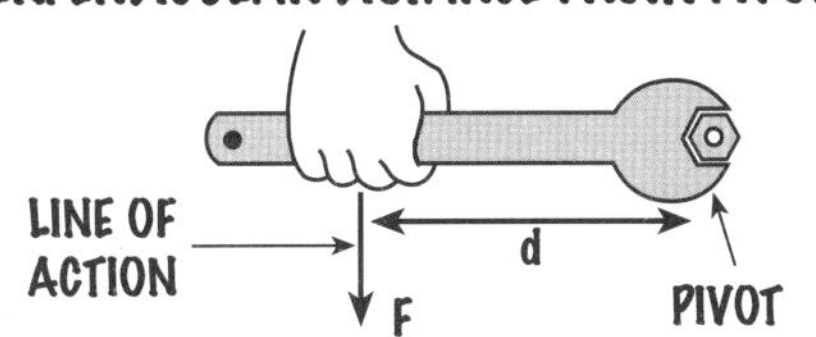

The relationship ...

MOMENT (Nm) = FORCE (N) x PERPENDICULAR DISTANCE FROM PIVOT (m)

Remember! The standard unit is the NEWTON - METRE.

Principle of Moments - balanced moments

ANTI-CLOCKWISE MOMENT (A-C.M) — CLOCKWISE MOMENT (C.M.) ... and when balanced, **TOTAL CLOCKWISE MOMENT = TOTAL ANTICLOCKWISE MOMENT**

Examples:

1) Calculate the unknown force if the see-saw below is balanced

A-C.M — 2m — 4m — C.M; 10N, F

When balanced, TOTAL C.M = TOTAL A-C.M

$F \times 4m = 10N \times 2m$

$4F = 20$

$F = \frac{20}{4} = 5N$

BOTH THESE FORMULAE ARE GIVEN

2) Two girls weighing 250N and 300N sit on one side of a seesaw 1m and 1.5m from the pivot respectively. Where must a boy weighing 350N sit in order to balance the seesaw?

A diagram is essential.

A-C.M — d — 1.5m — 1m — C.M; 350N, 250N, 300N

When balanced, TOTAL C.M = TOTAL A-C.M

$(250N \times 1m) + (300N \times 1.5m) = 350N \times d$

$250 + 450 = 350d$

$700 = 350d \quad d = \frac{700}{350} = 2m$

EFFECT OF APPLIED FORCES ON SOLID OBJECTS - HOOKE'S LAW

APPLIED FORCES acting on SOLID OBJECTS can result in ...
- EXTENSION ... where the object is STRETCHED.
- COMPRESSION ... where the object is SQUASHED.

Here we will consider in more detail the EXTENSION PRODUCED due to STRETCHING as the result of an APPLIED FORCE on an OBJECT e.g. METAL WIRE or SPRING. A typical graph of EXTENSION against APPLIED FORCE would look like:

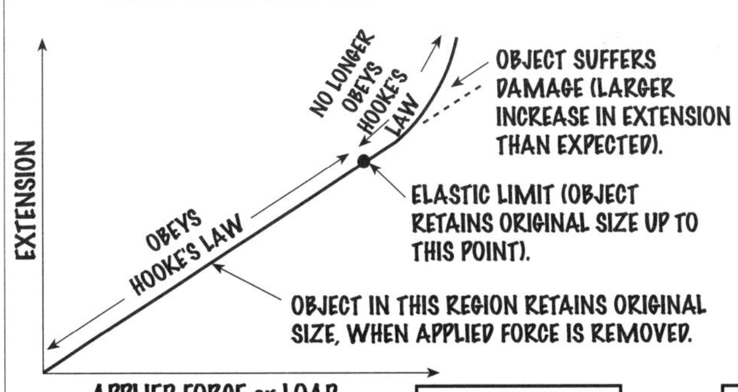

Hooke's Law

Up to the ELASTIC LIMIT ...
- ... EXTENSION IS DIRECTLY PROPORTIONAL TO APPLIED FORCE ...
- ... i.e. extension is doubled if we double the applied force etc.
- This relationship between EXTENSION and
- APPLIED FORCE ...
- ... is known as HOOKE'S LAW.

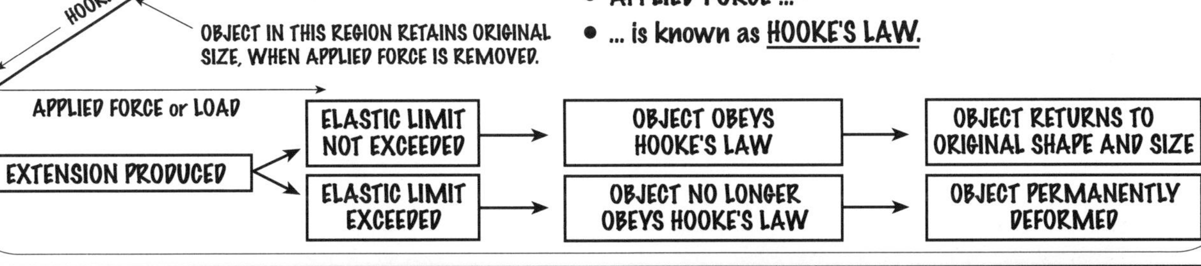

KEY POINTS:

- Moment (Nm) = Force (N) x Perpendicular distance from pivot (m).
- For balanced moments, Total clockwise moment = Total anticlockwise moment
- If elastic limit is not exceeded a stretched object obeys Hooke's Law.

FORCE AND PRESSURE ON SOLIDS AND LIQUIDS

PRESSURE - FORCE ACTING ON UNIT AREA OF A SURFACE

When a **FORCE** acts on a **SURFACE** the **PRESSURE** exerted on that surface depends on

- the **FORCE THAT ACTS** - greater force, greater pressure.
- **AREA OF SURFACE** - greater area, smaller pressure.

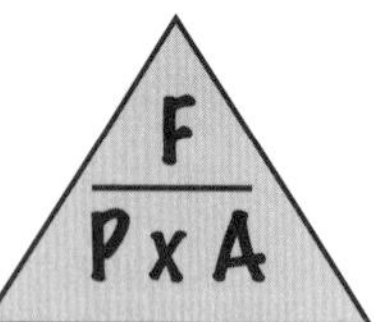

Pressure, force and area are related by the relationship.

$$\text{PRESSURE (N/m}^2) = \frac{\text{FORCE (N)}}{\text{AREA (m}^2)}$$

YOU MUST KNOW THIS FORMULA - SO LEARN IT.

- the **UNITS** are **NEWTONS PER METRE SQUARED**, N/m^2 also called **PASCALS, Pa.**
- Pressure may also be measured in **NEWTONS PER CENTIMETRE SQUARED**, N/cm^2.

Example

A box of weight 100N rests on a surface as shown here. Calculate a) the greatest. b) the least pressure that it can exert on the surface.

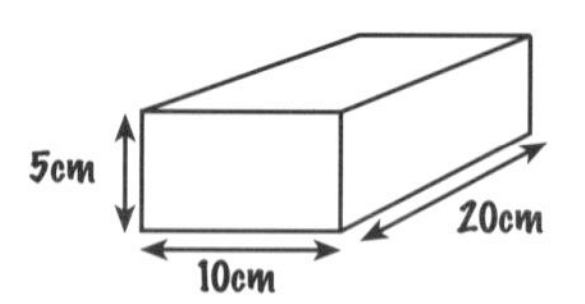

TWO VERY IMPORTANT FACTS.

1) The box can rest on surface in 3 ways • END ON • SIDEWAYS • FLAT
In each case the **AREA IN CONTACT IS DIFFERENT.**

2) The **WEIGHT** of the box does not change (regardless of position of box).

THIS UNIT DOES NOT EQUAL THE PASCAL

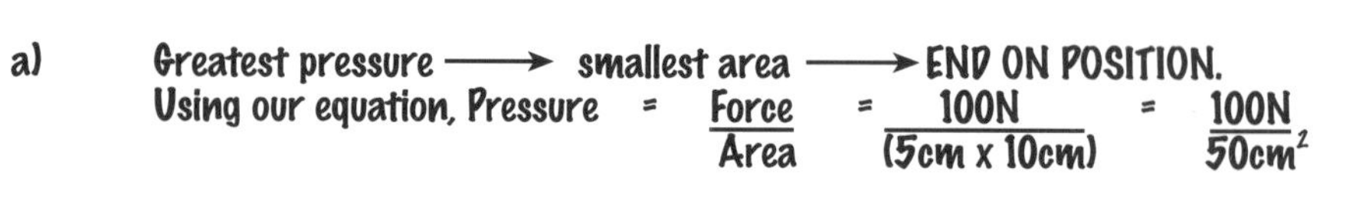

a) Greatest pressure → smallest area → END ON POSITION.
Using our equation, Pressure $= \frac{\text{Force}}{\text{Area}} = \frac{100\text{N}}{(5\text{cm} \times 10\text{cm})} = \frac{100\text{N}}{50\text{cm}^2} = 2\text{N/cm}^2$

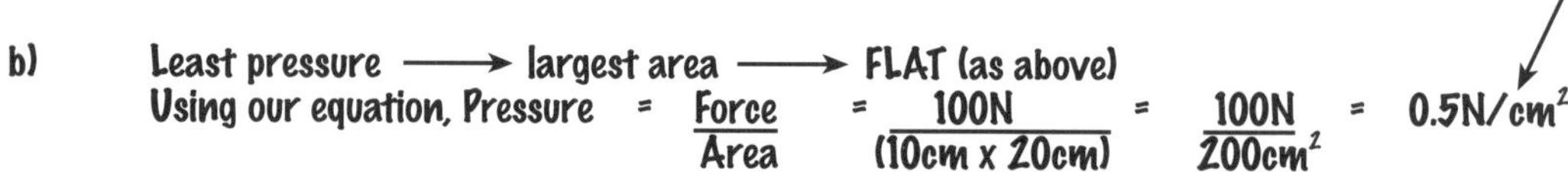

b) Least pressure → largest area → FLAT (as above)
Using our equation, Pressure $= \frac{\text{Force}}{\text{Area}} = \frac{100\text{N}}{(10\text{cm} \times 20\text{cm})} = \frac{100\text{N}}{200\text{cm}^2} = 0.5\text{N/cm}^2$

APPLICATIONS OF PRESSURE

The amount of **PRESSURE EXERTED** on a **SURFACE** has some important practical applications in various situations.

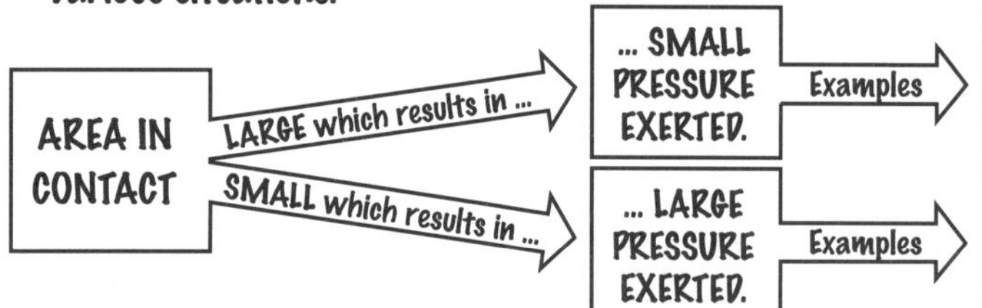

1. FOUNDATION OF BUILDINGS
- To prevent building SINKING INTO THE GROUND.
2. SNOW SHOES AND SKIS
- To prevent SINKING INTO THE SNOW.

1. ICE SKATES
- Causes ICE to MELT at AREA of CONTACT, easier to skate.
2. SHARP KNIFE BLADES
- Makes it easier to 'CUT THROUGH' food.

PRESSURE IN LIQUIDS

Very simply: **THE PRESSURE IN A LIQUID INCREASES WITH DEPTH OF LIQUID**

This little demonstration shows this perfectly ...

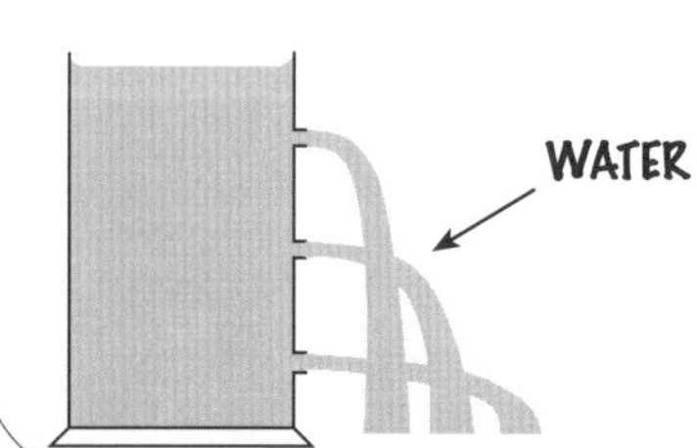

THE DEEPER THE HOLE THE GREATER THE FORCE ON THE WATER, THE FURTHER IT SPURTS OUT.

A practical application to counter this ...

WATER

DAM

CROSS-SECTIONAL AREA OF DAM INCREASES WITH DEPTH OF WATER TO COUNTER THE INCREASED PRESSURE

KEY POINTS:

- Pressure = Force ÷ Area. • Measured in N/cm^2 or N/m^2 (Pa).
- The pressure in a liquid increases with depth.

FORCE AND PRESSURE ON LIQUIDS AND GASES

PRESSURE IN LIQUIDS - HYDRAULIC SYSTEMS

Any hydraulic system depends on two principles

- LIQUIDS ARE INCOMPRESSIBLE (cannot be squashed)
- AT ANY POINT WITHIN THE SYSTEM, THE PRESSURE IS ALWAYS THE SAME.

What are they used for?

- ... to DIRECT FORCE to where it is required ...
- ... and as FORCE MULTIPLIERS.

How they work

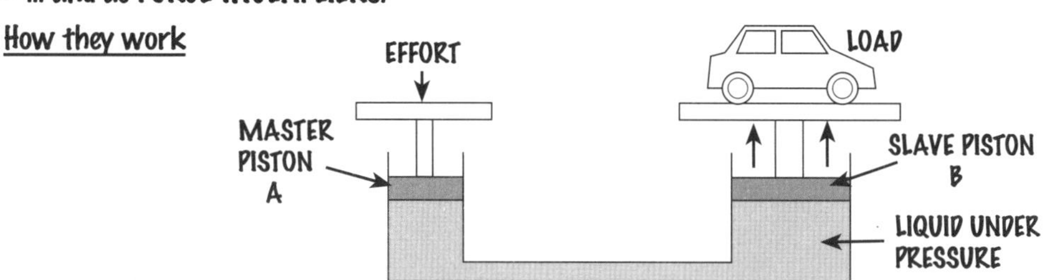

In the above diagram, ...

- EFFORT FORCE INCREASES PRESSURE IN LIQUID.
- PRESSURE is transmitted through system ...
- ... but AREA of B is greater than AREA of A.

Since FORCE = PRESSURE x AREA, the FORCE in B is MULTIPLIED!

EVERYDAY APPLICATIONS OF HYDRAULIC SYSTEMS

1) HYDRAULIC CAR BRAKES - the CLASSIC EXAMPLE!

- Brake pedal pushes MASTER PISTON ...
- ... exerting force on BRAKE FLUID.
- Pressure is transmitted to SLAVE PISTONS on wheel discs ...
- ... which MULTIPLY INITIAL FORCE ...
- ... causing BRAKE PADS to be ...
- ... forced against the wheels.

2) HYDRAULIC LIFTING DEVICES (see above)

GAS PRESSURE

Consider the following situation for a FIXED MASS OF GAS (amount of gas does not change) at CONSTANT TEMPERATURE. The gas is enclosed in a GAS SYRINGE connected to a suitable PRESSURE GAUGE.

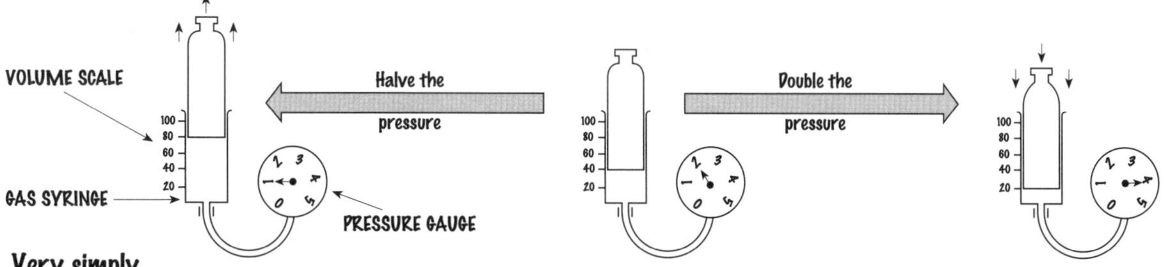

Very simply

- As we INCREASE the PRESSURE on a gas its VOLUME DECREASES ...
- ... i.e. VOLUME IS INVERSELY PROPORTIONAL TO PRESSURE, providing the TEMP has NOT CHANGED.

KEY POINTS:

- Hydraulic systems transmit pressure and act as force multipliers.
- Volume is inversely proportional to pressure, providing the temp. and mass of gas hasn't changed.

THE SOLAR SYSTEM I

All BODIES including the SUN, EARTH, MOON and other PLANETS attract each other with a FORCE called GRAVITY. This GRAVITATIONAL FORCE between two bodies is the main factor which controls the movement of a small body around a larger one.

THE PLANETS

- The planets (including Earth) are NON LUMINOUS BODIES ...
  ... which orbit STARS (in our case the SUN).
- We see the OTHER PLANETS because ...
  ... LIGHT from the SUN, REFLECTS off them.
- The ORBITS of planets are ELLIPTICAL (slightly squashed circles) ...
  ... in the SAME PLANE (except Pluto) with the SUN at the centre.

THE SUN

- The SUN is a STAR ...
  ... and like all stars ...
  ... is a SOURCE of LIGHT ...
  ... and other forms of ...
  ... ELECTROMAGNETIC RADIATION.

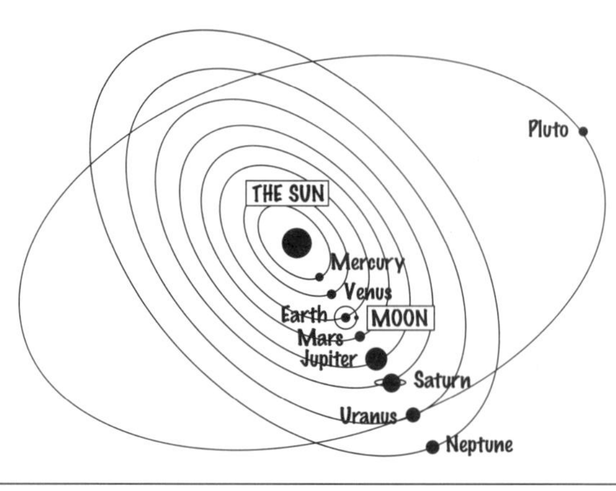

MOONS

- These are BODIES that
  ... ORBIT the PLANETS.
- Not all planets have moons ...
  ... but the EARTH has ONE ...
  ... called the MOON!!

N.B.
YOU NEED TO KNOW THE ORDER OF THE PLANETS FROM THE SUN OUTWARDS.

THE PLANETS ARE NATURAL SATELLITES OF THE SUN WHILE MOONS ARE NATURAL SATELLITES OF THE PLANETS.

THE EARTH

SUN LIGHT N AXIS DARK S

- Spins on a TILTED AXIS ...
  ... completing ONE REVOLUTION EVERY 24 HOURS, giving us DAY and NIGHT ...
  ... which explains the APPARENT MOVEMENT OF THE STARS AT NIGHT ...
  ... REMEMBER, it isn't THE STARS THAT ARE MOVING AROUND BUT THE EARTH, SPINNING ON ITS AXIS!
- Completes ONE ORBIT of the SUN every 365 ¼ days (an EARTH YEAR) giving us the DIFFERENT SEASONS.

NORTHERN HEMISPHERE SUMMER

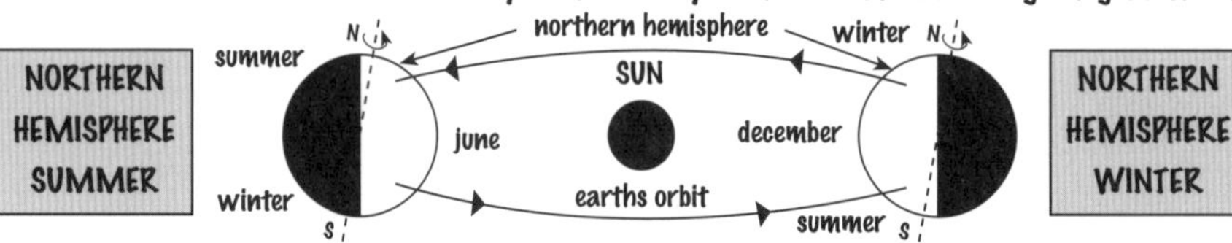

NORTHERN HEMISPHERE WINTER

- Northern Hemisphere tilted TOWARDS THE SUN ...
- ... so DAYTIME IS LONGER THAN NIGHT TIME ...
- ... and SUN RISES HIGHER IN THE SKY ...
- ... delivering MORE ENERGY to this hemisphere ...
- ... which makes it WARMER (hopefully!)

- Northern Hemisphere tilted AWAY FROM SUN ...
- ... so NIGHT TIME IS LONGER THAN DAY TIME ...
- ... and SUN IS LOWER IN SKY ...
- ... and so delivers LESS ENERGY to this hemisphere ...
- ... which makes it COLDER.

GRAVITATIONAL FORCE BETWEEN TWO BODIES - inverse square relationship

Consider TWO BODIES a DISTANCE 'D' apart and the GRAVITATIONAL FORCE between them is 'F' then ...

- ... if their distance apart becomes '2D' (TWICE ORIGINAL) ...
  - ... the gravitational force between the two bodies becomes $\frac{F}{4}$ (ONE QUARTER ORIGINAL!) and ...
- ... if their distance apart becomes '3D' (THREE TIMES ORIGINAL)
  - ... then gravitational force between the two bodies becomes $\frac{F}{9}$ (ONE NINTH ORIGINAL!)

and so on ...

KEY POINTS:

- The planets in order from the Sun are: Mercury, Venus, Earth, Mars, Jupiter, Saturn, Uranus, Neptune and Pluto.
- The gravitational force between two bodies obeys the inverse square relationship.

ORGANISATION OF THE SOLAR SYSTEM - THE THEORIES

For over 2000 years there have been a variety of different ideas concerning the arrangement of the solar system and beyond. The acceptance or rejection of any particular idea has depended on the social and historical context in which it was developed and proposed. You are not expected to recall specific ideas.

PYTHAGORAS
- EARTH at the centre with ... ... all the stars in orbit around it.
- Called the GEOCENTRIC model and was ... ... accepted by the rulers of ancient Greece.

PTOLEMY
- Proposed updated ... ... Geocentric model where ... Sun circled the Earth and ... ... other planets circled the Sun.

GALILEO GALILEI
- Took up on Copernicus' ideas and using the telescope ... ... discovered four moons going around Jupiter ... ... which was evidence for the heliocentric model.
- Again the church was not impressed and he was jailed for life.

YEARS B.C. 600 400 200 0 200 400 600 800 1000 1200 1400 1600 1800 YEARS A.D.

THALES
- EARTH is a ... ... disc floating ... ... on water.

ARISTARCHUS
- SUN at the centre with ... ... Earth in orbit around it.
- Called the HELIOCENTRIC model but ... ... rejected as he had no experimental support.

NICOLAUS COPERNICUS
- Improved on Aristarchus' idea and provided ... ... experimental support for the heliocentric model.
- Ideas not published until close to his death as they ... ... would challenge the beliefs of the authorities and church.

ISAAC NEWTON
- His theory of gravity using ... ... the ideas of Galileo ... ... finally proved that ... ... Copernicus was correct after all.

ARTIFICIAL SATELLITES - in orbit around the Earth

They have many uses:

1. OBSERVATION OF THE EARTH
- Military purposes for 'spying'... ... capable of taking pictures of minute detail.
- Take photographs of 'disasters' ... ... floods, earthquakes, crop failures.

2. WEATHER MONITORING
- Weather satellites have a ... ... low polar orbit over North and South pole.
- They collect information about the atmosphere ... ... including cloud photographs and monitor their ... ... movement so weather forecasts can be made.

3. EXPLORATION OF THE SOLAR SYSTEM
- Space telescope (e.g. Hubble) ... ... orbit above the Earth's atmosphere.
- Solar system and beyond observed with no ... ... interference as atmosphere absorbs and scatters light ... ... while clouds and weather storms ... ... also interfere with the light.

4. COMMUNICATIONS SYSTEMS
- Radio, TV, Telephone links ... ... for places far apart.
- This is a geosynchrous satellite ... ... as it moves at exactly the same rate ... ... as the Earth revolves (takes 24 hrs to go around).
- Remains at same position when viewed from earth.

ASTEROIDS AND COMETS

ASTEROIDS are a band of rock debris which occupy a belt between the orbits of MARS and JUPITER while ...

COMETS ...
- have a CORE OF FROZEN GAS and DUST ... ... and an ELLIPTICAL ORBIT around the sun.

WHICH IS IN A DIFFERENT PLANE FROM THAT OF THE PLANETS.

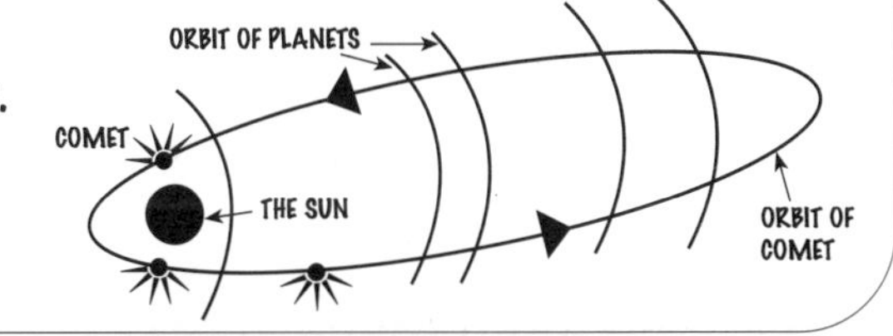

KEY POINTS:
- Two theories about the organisation of the solar system have been the Geocentric model and the Heliocentric model.

THE UNIVERSE

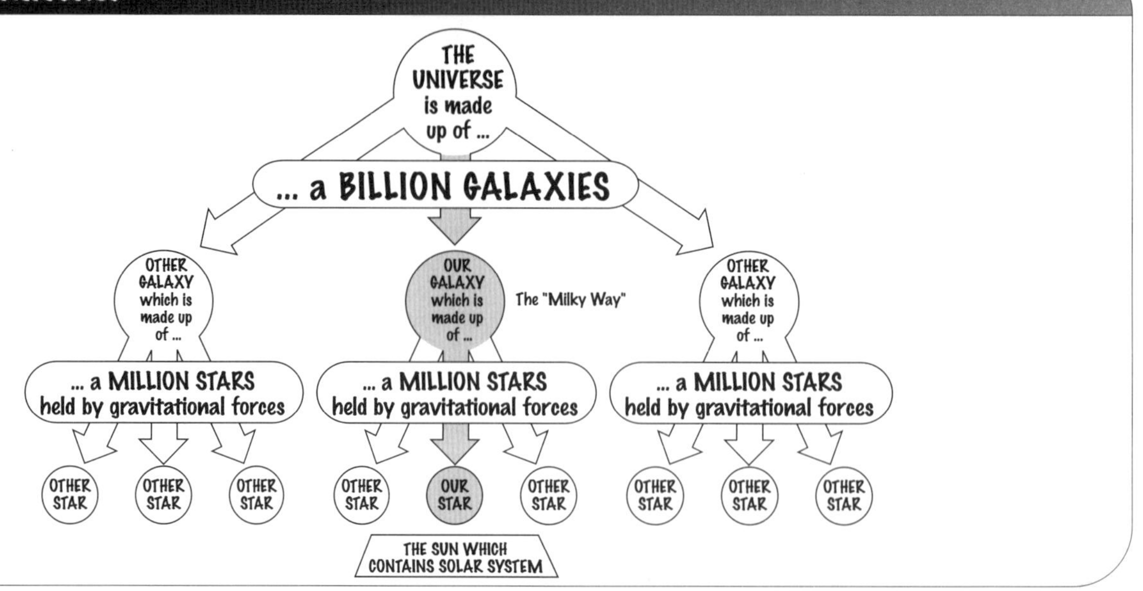

MEASURING THE UNIVERSE - LIGHT YEARS

A LIGHT YEAR is the DISTANCE travelled by LIGHT in ONE EARTH YEAR ...

- ... and is equal to 9,500 000 000 000 km!!! (light does travel at 300,000,000m/s - see W.4).

We use this unit as the distances involved are so enormous and mind blowing!...

- Our nearest star (ALPHA CENTAURI) is 4.3 LIGHT YEARS away while ...
- ... our nearest galaxy (ANDROMEDA) is a mere 2,200,000 LIGHT YEARS away!!

EVOLUTION OF THE UNIVERSE - THE BIG BANG THEORY

Very simply ...

- The UNIVERSE started at ONE PLACE BILLIONS OF YEARS AGO with a HUGE EXPLOSION ...
- ... and is constantly EXPANDING because ...
- ... distant galaxies are moving further and further from each other although ...
- ... the expansion is slowing down which means ...

UNIVERSE WILL KEEP ON EXPANDING FOR EVER AND EVER ... or ... GRAVITY WILL TAKE ITS EFFECT CAUSING THE UNIVERSE TO STOP EXPANDING AND TO START COLLAPSING

EVOLUTION OF STARS

STAGE 1: A STAR IS BORN (sorry)

- STARS, including our sun, are formed when ...
- ... CLOUDS of DUST and GAS are pulled together by ...
- ... GRAVITATIONAL ATTRACTION between the particles.
- As the cloud SHRINKS it becomes HOTTER and HOTTER, to become a STAR.

cloud of dust and gas → STAR

STAGE 2: A STABLE LIFE (our sun is at this stage!)

- The MASSIVE FORCES OF ATTRACTION pulling INWARDS are BALANCED by FORCES ACTING OUTWARDS ...
- ... which are caused by the HUGE TEMPERATURES within the star.

STAGE 3: DEATH OF A STAR

STAR → RED GIANT → WHITE DWARF

- SUPPLY OF HYDROGEN RUNS OUT ...
- ... and the star SWELLS UP becoming ...
- ... COLDER and REDDER.

- Continues to cool down and will eventually COLLAPSE under ...
- ... its own GRAVITY resulting in material which is ...
- ... MILLIONS OF TIMES DENSER THAN ANY MATERIAL ON EARTH!!

KEY POINTS:

- The Universe is made up of a billion galaxies with each galaxy containing a million stars.
- A light year is the distance travelled by light in one earth year.

1. What is speed? Name three units.
2. A boy runs 1200m in 5 minutes. What is his average speed?
3. What does the gradient of a distance-time graph represent?
4. What is a force?
5. What are forces measured in?
6. What are a) balanced b) unbalanced forces?
7. What effect do balanced and unbalanced forces have on a) a stationery object b) an object already moving?
8. What is friction?
9. When does friction exist?
10. Which four factors affect the stopping distance of a vehicle?
11. What is acceleration?
12. A cyclist takes 20 seconds to reach 10m/s from rest. What is the acceleration?
13. A car travelling at 20m/s increased its speed to 35m/s in 5 seconds. What is the acceleration?
14. What is the formula relating force, mass and acceleration?
15. What is mass?
16. What is weight?
17. What is gravity?
18. What is the name of the forces which act on a falling object?
19. If an object reaches terminal velocity, which two forces are balanced?
20. What is a moment?
21. What is the principle of moments?
22. Draw a labelled diagram of extension against applied force for an elastic object.
23. What is the elastic limit?
24. Draw a graph of extension against applied force for a material using the following results.

Applied Force (N)	0	10	20	30	40	50	60	70
Extension (mm)	0	5	10	15	20	25	31	39

25. On your graph mark in a) region where material retains original size b) elastic limit c) region material suffers permanent damage.
26. A girl weighing 500N stands on one foot of area 40cm^2. What pressure does she exert on the ground?
27. What is the simple relationship between pressure in a liquid and depth of liquid?
28. On which 2 principles do hydraulic systems depend?
29. Name 2 applications of hydraulic systems.
30. What is the relationship between the pressure and volume of a gas at constant temperature? What else must be constant?
31. What is the Sun?
32. What is the order of the planets from the Sun outwards?
33. What is the inverse square relationship for any two bodies?
34. Name any two uses for artificial satellites.
35. What are asteroids?
36. What are comets?
37. What is the Universe made up of?
38. What is a light year?
39. What is the big bang theory?
40. How are stars a) born b) die?

ENERGY RESOURCES AND ENERGY TRANSFER

ENERGY RESOURCES I - Non-renewables E 1

- NON-RENEWABLE ENERGY RESOURCES are those that will ONE DAY RUN OUT and ...
- ... once they have been used they CANNOT BE USED AGAIN.

THE OPTIONS:-

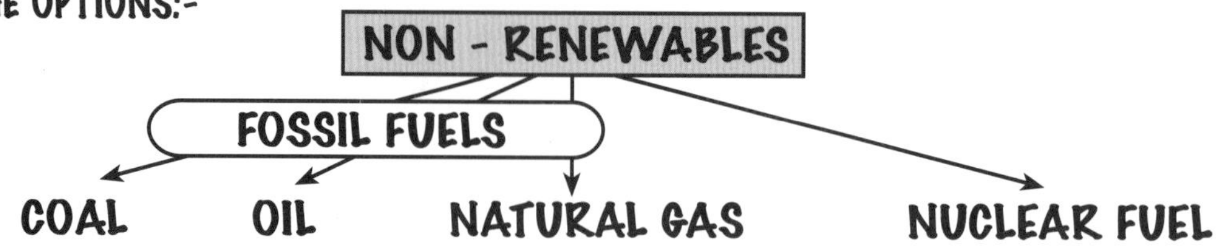

They are ALL used in the GENERATION of ELECTRICITY where ...
... THE **FUEL** IS USED TO GENERATE **HEAT**, WHICH THEN BOILS WATER TO MAKE **STEAM** TO DRIVE THE **TURBINES**, WHICH TURN THE **GENERATORS** PRODUCING **ELECTRICITY**.

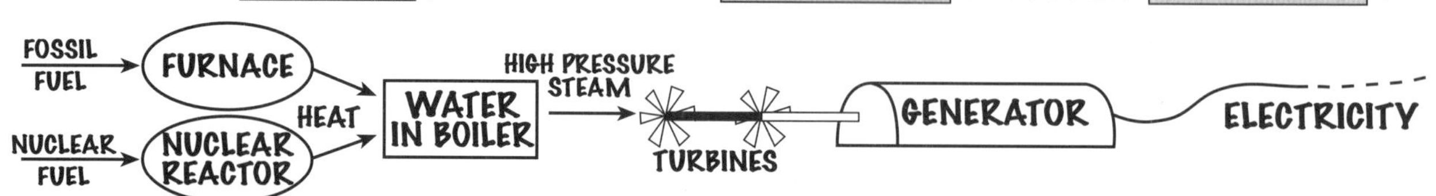

CONSERVATION OF NON - RENEWABLE ENERGY RESOURCES

Over 90% of the electricity produced in the world is produced by using non-renewable resources.
There is a growing awareness that CONSERVATION of these resources is needed for many reasons:-

	COAL (FOSSIL FUEL)	OIL/GAS (FOSSIL FUELS)	NUCLEAR (URANIUM + PLUTONIUM)
FINITE NATURE	ONLY 100 YRS. WORTH OF COAL LEFT.	PERHAPS 30 YRS. OF OIL/GAS LEFT.	GOOD SUPPLIES BUT IS DIFFICULT TO OBTAIN.
ENVIRONMENTAL PROBLEMS	GLOBAL WARMING, ACID RAIN (CAN BE REDUCED BY REMOVING SULPHUR DIOXIDE AFTER BURNING, CALLED 'SCRUBBING')	GLOBAL WARMING TANKER SPILLAGE	RADIOACTIVE SUBSTANCES MAY ESCAPE (SITE THE POWER STATION IN 'REMOTE' LOCATIONS WITH MANY FAIL SAFE DEVICES)
ETHICAL CONSIDERATIONS	GLOBAL WARMING IS CAUSING AN INCREASE IN THE TEMP. OF THE EARTH. ACID RAIN CAUSES DAMAGE TO TREES, PLANTS, FISH AND BUILDINGS.	GLOBAL WARMING IS CAUSING AN INCREASE IN THE TEMP. OF THE EARTH. ANY TANKER SPILLAGE HAS A DEVASTATING EFFECT ON MARINE LIFE.	DISCHARGE OF RADIOACTIVE LIQUIDS INTO THE SEA WILL AFFECT THE FOOD CHAIN. WASTE PRODUCTS NEED VERY SAFE STORAGE FOR THOUSANDS OF YEARS.
CONCERNS FOR THE FUTURE	THERE IS A CONCERN THAT ALL PEOPLE, BOTH NOW AND IN THE FUTURE SHOULD HAVE A FAIR AND APPROPRIATE SHARE OF THE EARTH'S RESOURCES. APART FROM GENERATING ELECTRICITY, FOSSIL FUELS ESPECIALLY, HAVE MANY OTHER VALUABLE USES (MAKING OF PLASTICS etc) AND FOR TRANSPORT (OIL).		

KEY POINTS:

- Coal, Oil, Natural Gas (the fossil fuels) and Nuclear fuel are non-renewable energy resources.

ENERGY RESOURCES II - Renewables

- RENEWABLE ENERGY RESOURCES are those that will NOT RUN OUT and ...
- ... are CONTINUALLY BEING REPLACED.

THE OPTIONS:-

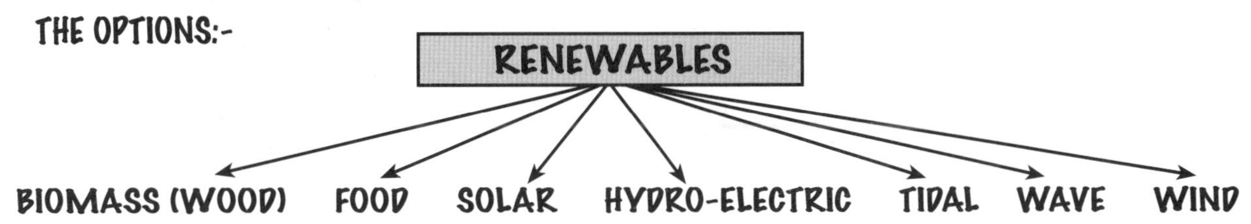

THESE CAN BE USED DIRECTLY ...

- PLANTS, ESPECIALLY TREES, CAN BE GROWN TO PROVIDE FUEL FOR HEATING.
- CROPS CAN BE GROWN AND HARVESTED TO PROVIDE FOOD SUPPLIES.

These are ALL used in the GENERATION of ELECTRICITY where ...
... THE ENERGY RESOURCE IS USED TO DRIVE TURBINES DIRECTLY etc ...
... WITH THE EXCEPTION OF SOLAR ...
... SO NO NASTY BURNING IS INVOLVED.

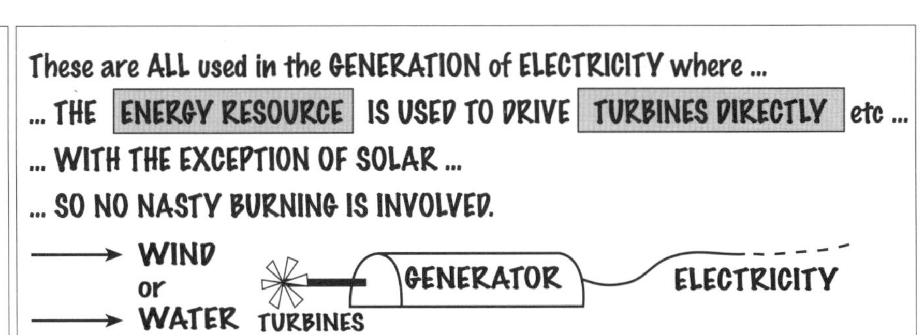

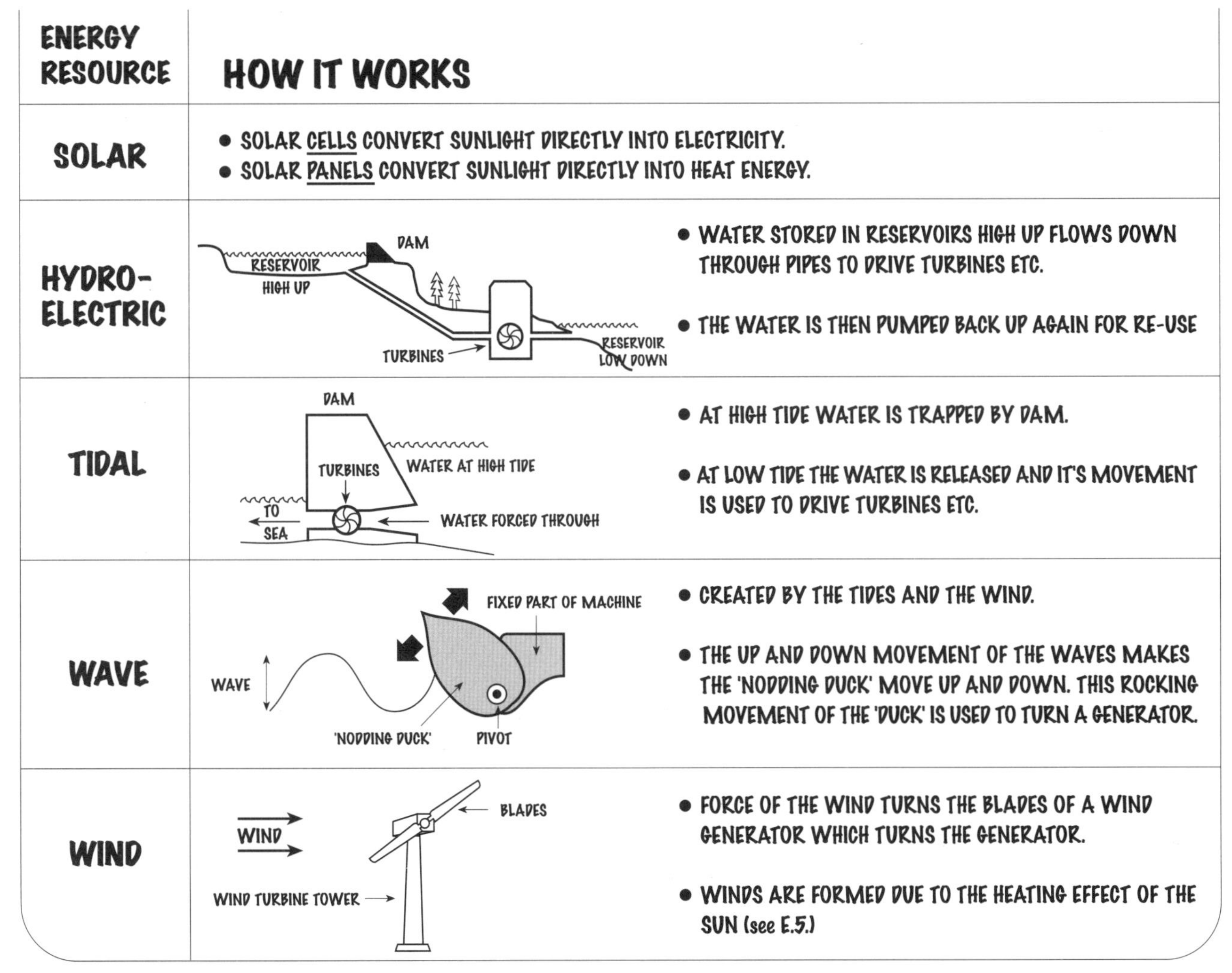

ENERGY RESOURCE	HOW IT WORKS
SOLAR	• SOLAR CELLS CONVERT SUNLIGHT DIRECTLY INTO ELECTRICITY. • SOLAR PANELS CONVERT SUNLIGHT DIRECTLY INTO HEAT ENERGY.
HYDRO-ELECTRIC	• WATER STORED IN RESERVOIRS HIGH UP FLOWS DOWN THROUGH PIPES TO DRIVE TURBINES ETC. • THE WATER IS THEN PUMPED BACK UP AGAIN FOR RE-USE
TIDAL	• AT HIGH TIDE WATER IS TRAPPED BY DAM. • AT LOW TIDE THE WATER IS RELEASED AND IT'S MOVEMENT IS USED TO DRIVE TURBINES ETC.
WAVE	• CREATED BY THE TIDES AND THE WIND. • THE UP AND DOWN MOVEMENT OF THE WAVES MAKES THE 'NODDING DUCK' MOVE UP AND DOWN. THIS ROCKING MOVEMENT OF THE 'DUCK' IS USED TO TURN A GENERATOR.
WIND	• FORCE OF THE WIND TURNS THE BLADES OF A WIND GENERATOR WHICH TURNS THE GENERATOR. • WINDS ARE FORMED DUE TO THE HEATING EFFECT OF THE SUN (see E.5.)

You'll notice all the non-renewables belt out energy but ...
... are environmentally disastrous as they are slowly poisioning the earth while ...
... the renewables are very 'earth friendly' but can't meet the demand. IS THERE A WINNER?

KEY POINTS:

- Biomass, Food, Solar, Hydro-electric, Tidal, Wave and Wind are renewable energy resources.

THE SUN'S ENERGY INPUT TO THE EARTH

The SUN is the ORIGINAL SOURCE of MOST of the EARTH'S ENERGY RESOURCES ...
... both NON - RENEWABLE and RENEWABLE

e.g. formation of coal

REMEMBER!! THIS PROCESS TOOK PLACE MILLIONS OF YEARS AGO - A VERY IMPORTANT FACT.

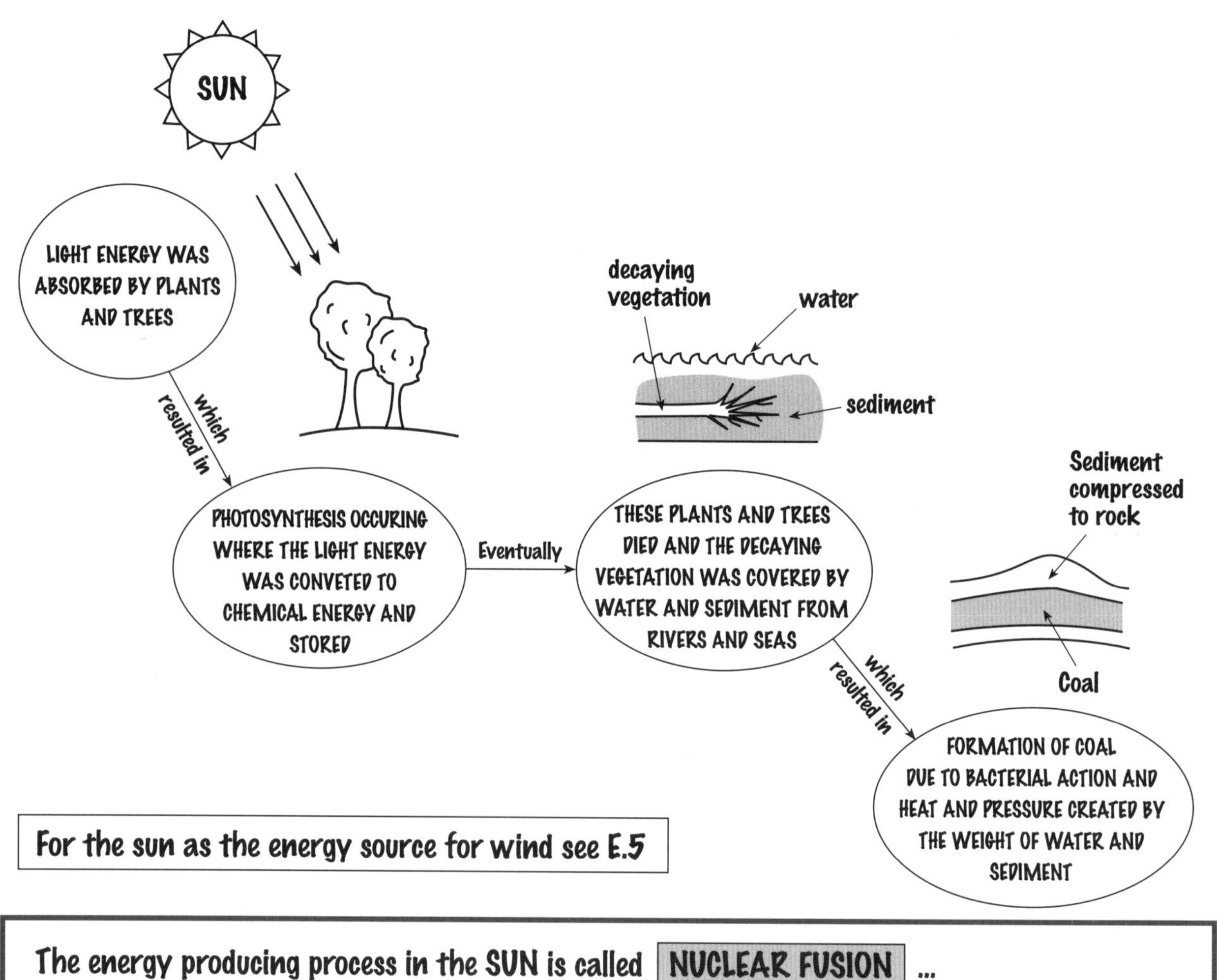

For the sun as the energy source for wind see E.5

The energy producing process in the SUN is called NUCLEAR FUSION ...

At centre of SUN, HYDROGEN nuclei at GREAT PRESSURE and VERY HIGH TEMP ... → ... Join together ... → ... to form HELIUM NUCLEI which have a SMALLER MASS than the HYDROGEN NUCLEI that FUSED TOGETHER to form them ... → ... so ... → ... this 'LOST MASS' is CONVERTED into HEAT ENERGY. Billions of these reactions happen every second.

KEY POINTS:

- The Sun is the original source of most non-renewable and renewable energy resources.
- The energy producing process in the Sun is called Nuclear Fusion.

TYPES OF ENERGY TRANSFER

ENERGY exists in many FORMS. ENERGY TRANSFERS involve the transfer of energy TO and FROM the following forms ...

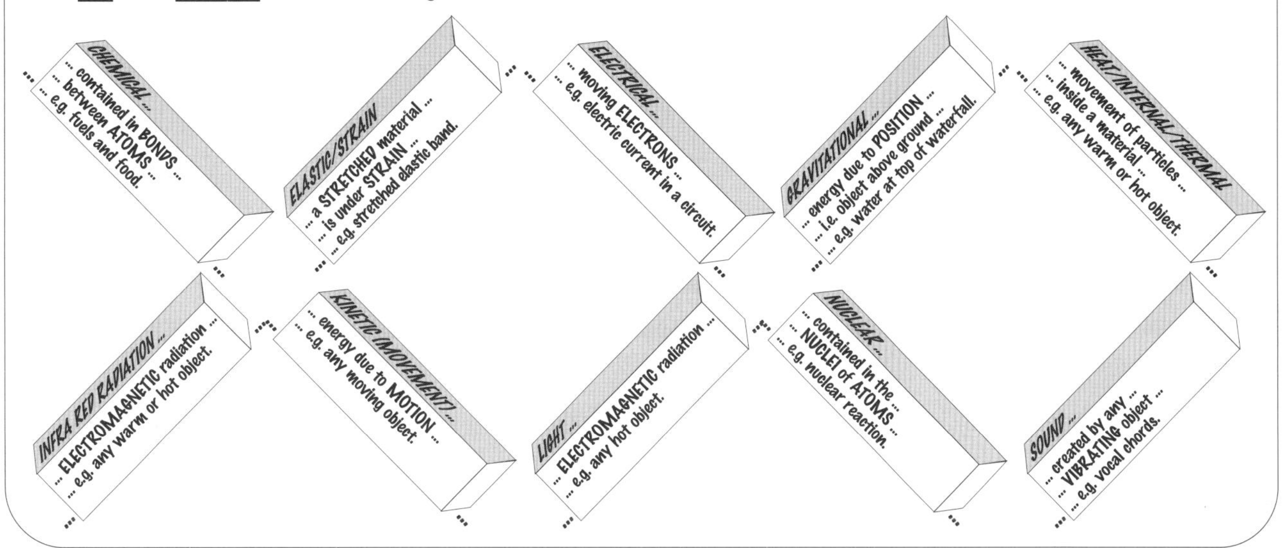

EFFICIENCY OF ENERGY TRANSFER

Whenever energy is transferred from one form to another some is 'wasted' - usually as heat and often sound, scientists are constantly working to reduce this wastage.
They want to make energy transfer devices more  EFFICIENT .
Here are FOUR examples of the intended energy transfer and wastage in everday devices:

1. Tungsten filament light bulb

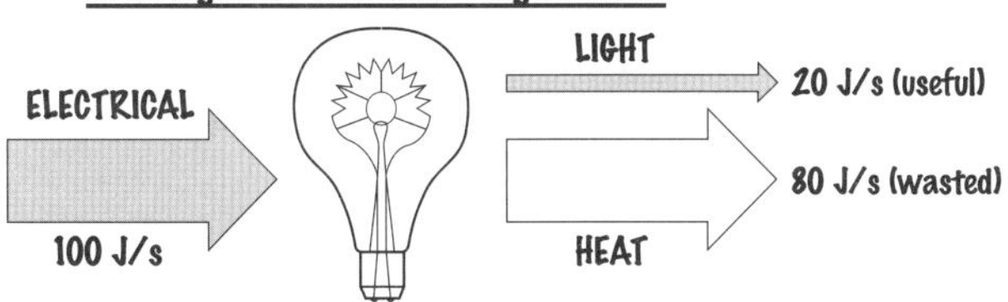

2. Low energy light bulb

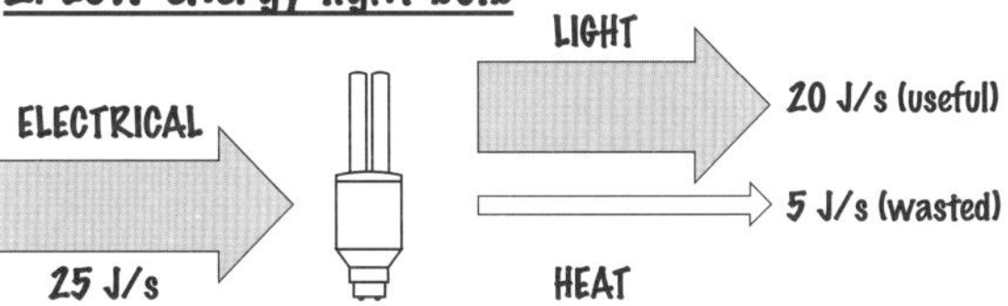

3. Electric kettle

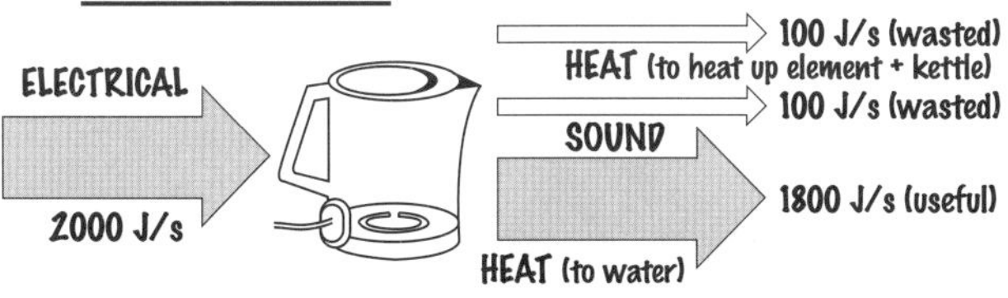

4. Electric motor (shown in a hairdrier)

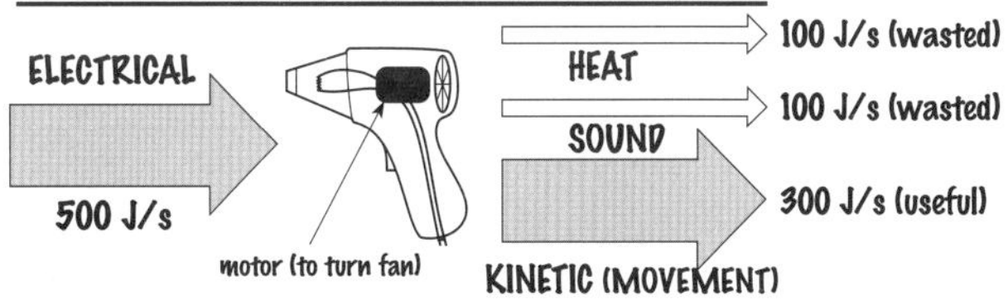

The EFFICIENCY of any transfer device is given by this relationship:

$$\text{EFFICIENCY} = \frac{\text{USEFUL ENERGY OUTPUT}}{\text{TOTAL ENERGY INPUT}}$$

So, for the filament light bulb above, Efficiency $= \frac{20}{100} \times 100 = 20\%$

and the low energy light bulb, Efficiency $= \frac{20}{25} \times 100 = 80\%$

(× 100: to convert it to a percentage)

... WHICH ONE WOULD YOU BUY?

You will be given this formula in the exam but you must know how to use it!!

KEY POINTS:

- Chemical, Infra red, Elastic, Kinetic, Electrical, Light, Gravitational, Nuclear, Heat and Sound are all forms of energy.
- Efficiency = Useful Energy Output ÷ Total Energy Input.

ENERGY TRANSFERS AND TEMPERATURE

Heat energy in Joules (J) is transferred from one place to another as a result of differences in temperature between the two places. In other words heat energy is transferred from hotter places to colder places.

CONDUCTION - Heat transfer in solids

- Heat is transferred in solids as a result of particles in the solid vibrating and passing on their movement energy to other nearby particles.
- Metals are all good conductors of heat energy because their electrons are free to move throughout the whole metallic structure which allows the heat energy to be passed through the metal quickly.
- Non-metallic substances including trapped air are poor conductors of heat energy.

CONVECTION - Heat transfer in liquids and gases

- Localised heating in a liquid or a gas causes particles in that region to move about more energetically. This causes them to move further apart and therefore become less dense than the rest of the substance. Because this region is now less dense it tends to rise through the rest of the fluid. It then cools and starts to fall again. This is a CONVECTION CURRENT and is responsible for ...

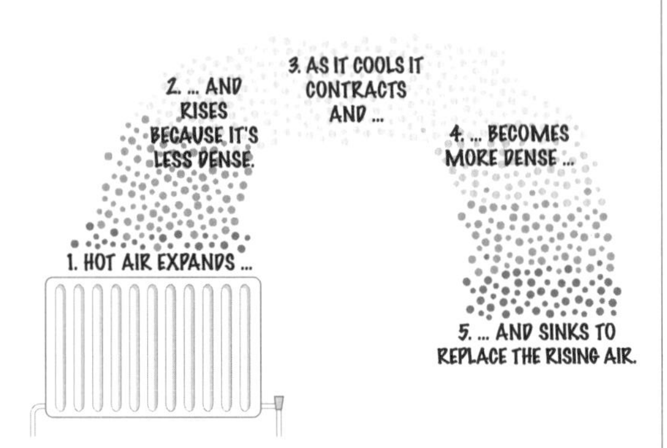

OCEAN CURRENTS

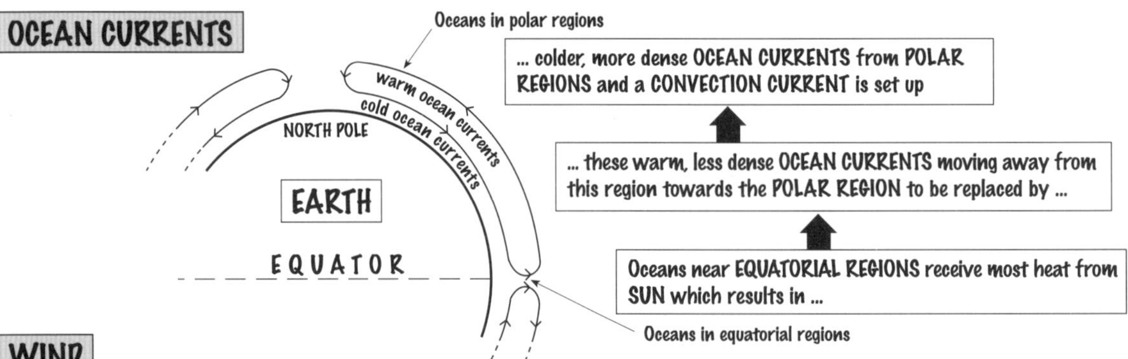

WIND

- Certain parts of the EARTH become warmer than other parts due to the HEAT ENERGY FROM THE SUN ...
- ... causing AIR above the warm parts to RISE, resulting in COOLER AIR from the colder parts ...
- ... moving into take its place.
- This is WIND and causes LAND and SEA BREEZES - the classic examples!

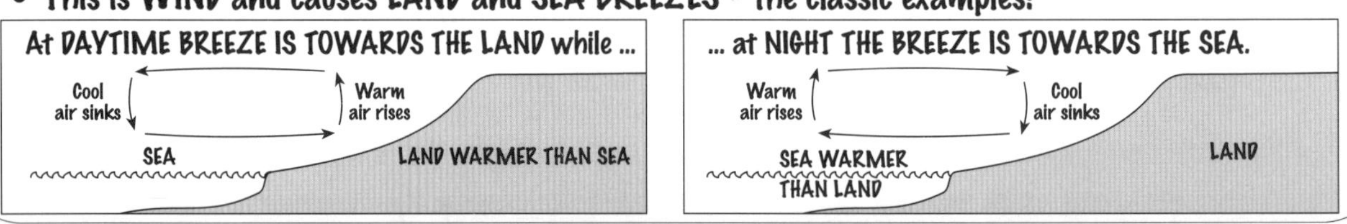

KEY POINTS:

- Heat is transferred through a solid by conduction.
- Heat is transferred through liquids and gases by convection.

3. RADIATION - Heat transfer through a vacuum!

- This is the transfer of energy by infra red radiation from hot objects. This radiation may pass through a vacuum and can be reflected by light shiny surfaces.
- The EMISSION and ABSORPTION of radiation depends on ...

1. SURFACE CONDITIONS

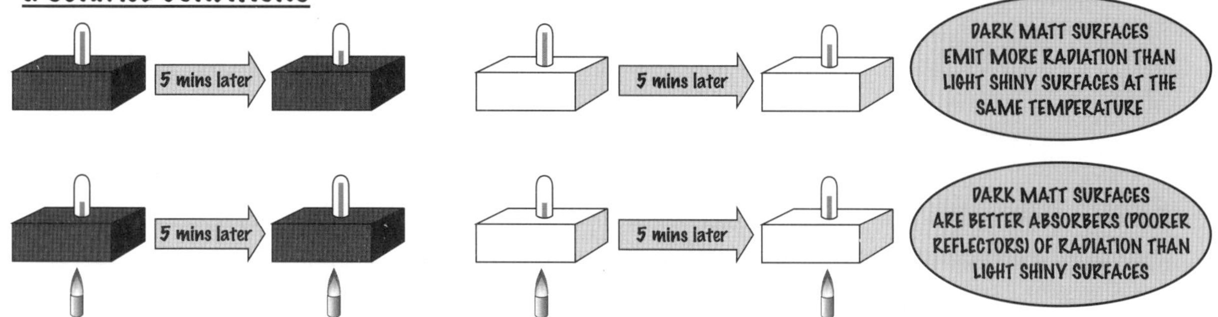

2. TEMPERATURE

If the temperature of the object is greater than its surroundings then it emits (gives out) radiation.
If the temperature of the object is less than its surroundings then it absorbs (takes in) radiation.

AMOUNT OF CARBON DIOXIDE IN THE ATMOSPHERE

CARBON DIOXIDE ABSORBS I-R RADIATION
... SO HEAT RADIATED BACK INTO SPACE BY EARTH IS ABSORBED BY CO_2 IN ATMOSPHERE.
IT IS THEN RE-RADIATED BACK TOWARDS THE EARTH.
THIS IS THE GREENHOUSE EFFECT AND IT'S EFFECT IS WORSENING WITH INCREASING CO_2.

4. EVAPORATION - Transfer of heat due to loss of particles from surface of liquid

RATE OF EVAPORATION OF WATER DEPENDS ON ...

1. SURFACE AREA OF WATER
Increase in surface area ... increases the evaporation ... ... as more water molecules are near the surface.

2. HUMIDITY OF AIR
Increase in the humidity decreases the evaporation ... ... as air above water already saturated with water molecules.

3. MOVEMENT OF AIR
Increased movement of air ... increases evaporation ... ... as any water molecules in air are 'blown away'.

4. TEMPERATURE
Increased temp. of water ... increases evaporation ... ... as more molecules have sufficient energy to escape.

- Liquid particles which have more energy than normal ...
- ... are able to escape from the surface of the liquid ...
- ... because they can overcome the forces of attraction ...
- ... of the less energetic liquid particles left behind.

KEY POINTS:

- Radiation is the transfer of energy from hot objects.
- Transfer of heat due to loss of particles from the surface of a liquid is evaporation.

REDUCING HEAT LOSSES

It is very important that any UNWANTED energy transfers from a HOT OBJECT to a COLDER OBJECT

... are kept to a MINIMUM as very simply ...

... ENERGY COSTS MONEY!!

One material which forms the basis of most energy saving devices is **AIR** !!!

1. THE HOUSE

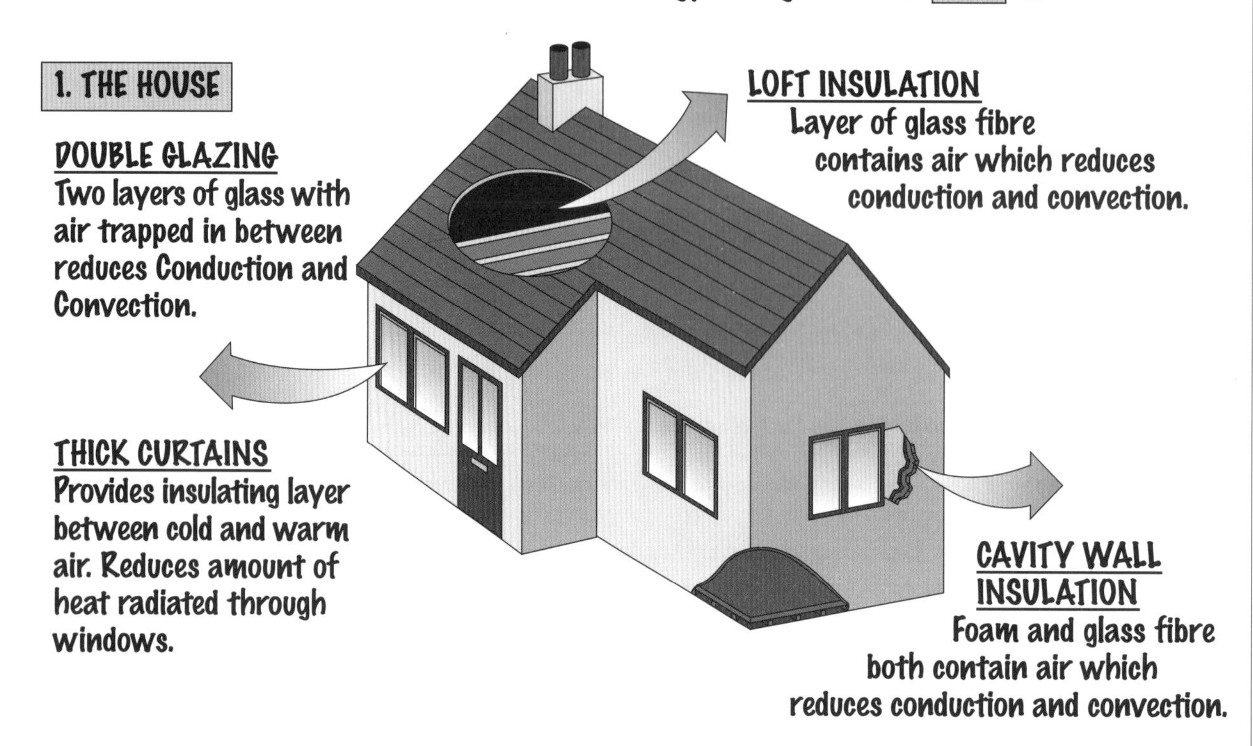

2. VACUUM FLASK

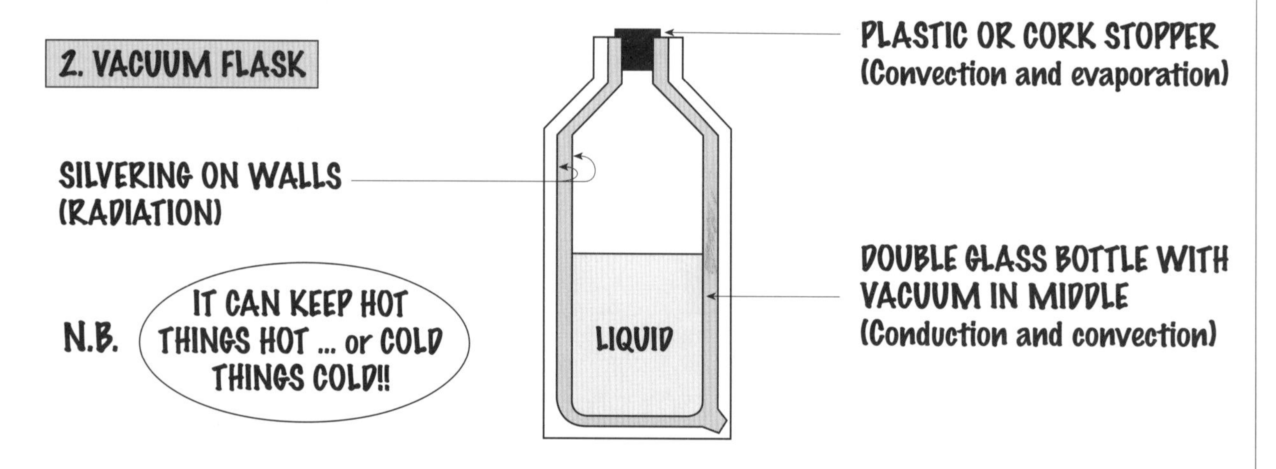

3. CLOTHING AND BEDDING

- Both made from materials that are themselves GOOD INSULATORS, however ...
- ... the insulating properties of each one is INCREASED ...
- ... because both also **TRAP AIR** within the material.

KEY POINTS:

- Air is a material which forms the basis of most energy saving devices.

WORK, POWER AND ENERGY I

- When a FORCE MOVES an OBJECT ...
- ... WORK is DONE ON THE OBJECT resulting in the TRANSFER OF ENERGY where ...

ENERGY TRANSFERRED (J) = WORK DONE (J) Both these are measured in JOULES.

Work done, force and distance moved are related by the relationship:

WORK DONE (J)	=	FORCE (N)	x	DISTANCE MOVED IN THE DIRECTION OF THE FORCE (m)

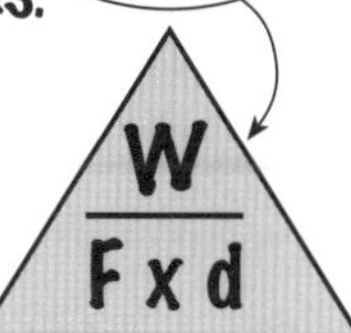

EXAMPLE

A man pushes a car for a distance of 20m at a constant velocity against a friction force of 250N. How much work does he do?

Using our formula: Work done = Force x distance moved
= 250 x 20
= 5,000J (or 5KJ)

i.e FORCE NEEDED TO OVERCOME FRICTION

POWER

- POWER is the RATE of TRANSFER OF ENERGY or ...
- ... the RATE at which WORK IS DONE.
- The GREATER the POWER ...
- ... the MORE ENERGY THAT IS TRANSFERRED EVERY SECOND.

The relationship:

$$\text{POWER (Watt, W or J/s)} = \frac{\text{WORK DONE (J)}}{\text{TIME TAKEN (s)}}$$

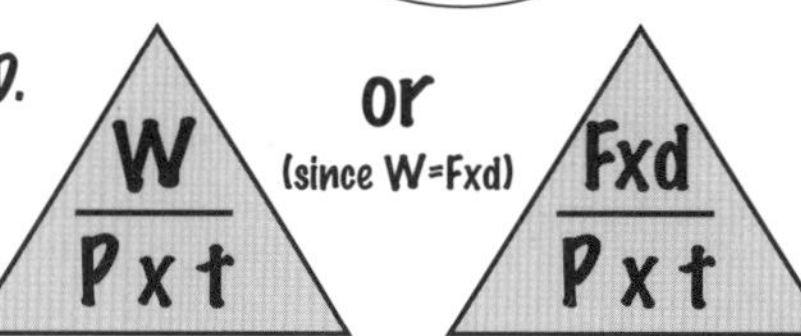

EXAMPLE 1. "A winch lifting a bucket of bricks transfers 300KJ of energy to the bucket in 15 seconds. Calculate the power of the winch."

NOTE. You've been given ...
- the ENERGY TRANSFERRED = WORK DONE AGAINST FORCE OF GRAVITY
- the TIME

But whatever you do look carefully at the units!

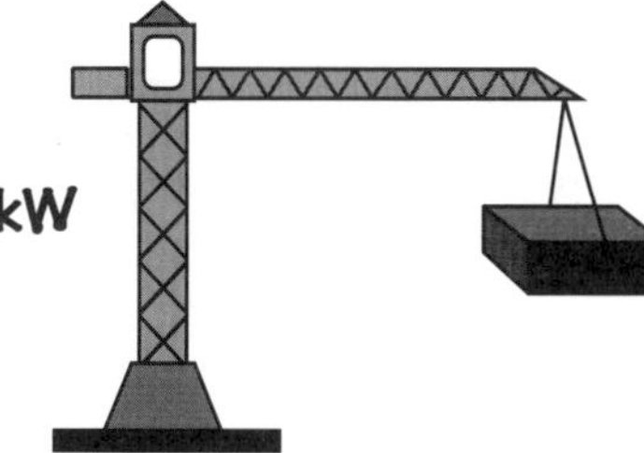

$$\text{POWER} = \frac{\text{WORK DONE}}{\text{TIME TAKEN}} = \frac{300KJ}{15s} = 20KJ/s$$ → This is kW

UNIT is KJ, ANSWER is KW. Answer = 20kW

(Don't worry about this! This means 20,000 Joules/sec, but 1000J = 1kJ, 20kJ/s = 20kWatts!! (or 20,000 Watts!!)

EXAMPLE 2. "A man lifts a load of 900N through a distance of 2m in a time period of 1.5sec. Calculate the power developed by him.

FIRSTLY WE NEED TO WORK OUT HOW MUCH WORK HE DOES (against gravity).

Using our formula: WORK DONE = FORCE x DISTANCE MOVED
= 900N x 2m
= 1,800J

i.e LOAD FORCE

Using our formula: $$\text{POWER} = \frac{\text{WORK DONE}}{\text{TIME TAKEN}} = \frac{1,800J}{1.5s} = 1,200J/sec$$ → This is Watts

Answer = 1,200Watts

KEY POINTS:

- Energy transferred (J) = Work done (J)
- Work done (J) = Force (N) x Distance moved in the direction of the force (m)
- Power (W or J/s) = Work done (J) ÷ Time taken (s)

KINETIC ENERGY

This is the ENERGY an object has because of its MOTION ...
... IF IT'S MOVING, IT'S GOT KINETIC ENERGY.

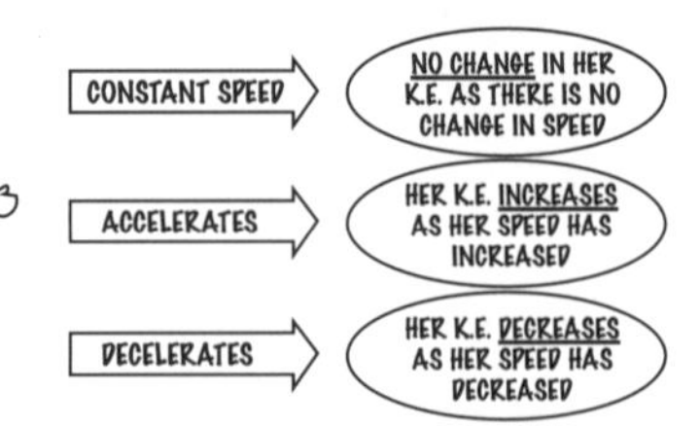

- We have already seen (see F.M.2) how increased speed increases the stopping distance of a vehicle. Since K.E. INCREASES with INCREASING SPEED, then its fairly obvious ...
  THE GREATER THE KINETIC ENERGY OF A VEHICLE ⟶ GREATER IT'S STOPPING DISTANCE

- It depends on two things:-
  - MASS OF OBJECT (Kg) - bigger its mass, bigger its K.E.
  - SPEED OF OBJECT (m/s) - bigger its speed, bigger its K.E.

The relationship:

$$\text{KINETIC ENERGY (J)} = \frac{1}{2} \times \text{MASS (Kg)} \times \text{(SPEED)}^2 \text{ (m/s)}^2$$

EXAMPLE

A car and it's driver has a total mass of 900 Kg and is travelling at a speed of 20m/s.
How much K.E. does it have?

Using our formula:

$$\text{K.E.} = \frac{1}{2} \times \text{MASS} \times \text{(SPEED)}^2$$
$$= \frac{1}{2} \times 900 \text{ Kg} \times (20\text{m/s})^2$$
$$= \frac{1}{2} \times 900 \times 400 = 180{,}000 \text{ Joules}$$

GRAVITATIONAL POTENTIAL ENERGY

This is the energy stored in an object due to it's position. If it has been lifted against the force of gravity, then it is capable of falling and therefore transferring this stored energy.

It depends on three things:-
- IT'S MASS (in kilograms)
- IT'S HEIGHT (in metres)

and
- THE ACCELERATION DUE TO GRAVITY (g) which is a CONSTANT as far as we are concerned and has a value of 10m/s^2 (it gets less, the further you go from earth).

The relationship:

$$\text{GRAVITATIONAL ENERGY (J)} = \text{MASS (Kg)} \times \text{ACCELERATION DUE TO GRAVITY } (\text{m/s}^2) \times \text{HEIGHT (m)}$$

One small but very important point to remember:
- the height must be the VERTICAL HEIGHT (i.e a straight line upwards!)

EXAMPLE

If an object has a mass of 10,000 Kg and is lifted to a height of 10m, calculate the gravitational potential energy transferred to the block.

GRAVITATIONAL POTENTIAL ENERGY = mass x g x HEIGHT
= 10,000Kg x 10m/s^2 x 10m
= 1,000,000 Joules (1,000kJ)

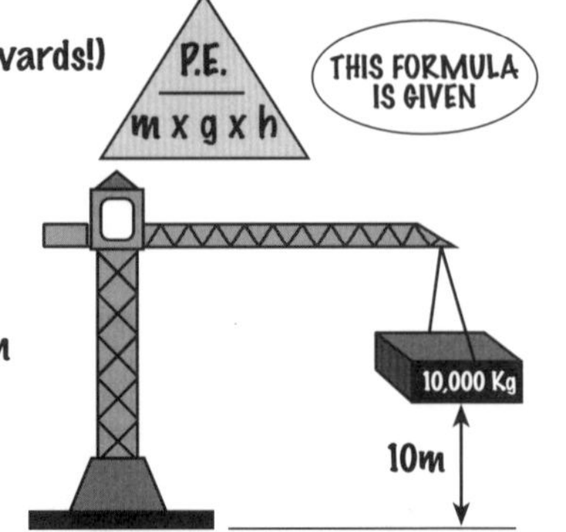

KEY POINTS:

- A moving object has Kinetic energy
- $KE = \frac{1}{2} \times \text{Mass} \times (\text{Speed})^2$
- The energy stored in an object due to its position is Gravitational potential energy.
- Gravitational energy = mass x g x height.

1. What are NON-RENEWABLE energy resources?
2. How are they used to generate electricity?
3. Why is it important to conserve these resources?
4. What are RENEWABLE energy resources?
5. Which resources are used for generating electricity?
6. Why are WOOD and FOOD called renewable energy resources?
7. Why is the SUN the ORIGINAL ENERGY RESOURCE?
8. How was COAL formed?
9. What is NUCLEAR FUSION?
10. Name TEN ENERGY FORMS.
11. Which ENERGY TRANSFERS take place when you switch on a TV?
12. Which ENERGY TRANSFERS take place when you switch a foodmixer on?
13. Write down the equation for EFFICIENCY.
14. For every 1000 Joules of chemical energy put into a car, only 240 Joules are transferred as movement energy. Calculate its EFFICIENCY.
15. What is temperature measured in?
16. Write down 4 main facts about transferring heat energy by CONDUCTION.
17. Write down 4 main facts about transferring heat energy by CONVECTION.
18. How are OCEAN CURRENTS and WIND created?
19. What is RADIATION?
20. How does a) temperature b) surface conditions and c) amount of carbon dioxide in the atmosphere affect the ABSORPTION and EMISSION of radiation from the earth?
21. What is EVAPORATION?
22. Write down as many ways as possible for preventing heat loss from a house by CONDUCTION.
23. Write down as many ways as possible for preventing heat loss from a house by CONVECTION.
24. How does a VACUUM FLASK keep hot things hot or cold things cold?
25. Why are CLOTHING and BEDDING good insulators?
26. What is the formula relating work, force and distance?
27. A car is pushed a distance of 120m against a friction force of 150N. How much work is done?
28. What is POWER?
29. What is the formula relating power, work and time?
30. What are the units of power?
31. A load of 1000N is lifted a distance of 3m in a time of 4 seconds. What is the power developed?
32. What is KINETIC ENERGY?
33. What two things does KINETIC ENERGY depend on?
34. If the kinetic energy of an object increases, then what must the object be doing?
35. What is the kinetic energy of a stationary object of mass 10kg?
36. What is the kinetic energy of a girl of mass 50kg running at a speed of 4m/s?
37. What is GRAVITATIONAL ENERGY?
38. What is the formula for gravitational energy?
39. A man of mass 80kg climbs up a ladder to a height of 6m. How much gravitational energy does he have?
40. An object has gravitational energy of 18000J. If it is all converted into kinetic energy, what is the speed of the object if it has a mass of 10kg?

ELECTRICITY AND MAGNETISM

ELECTRIC CHARGE 1 — EM 1

WHAT IS STATIC ELECTRICITY?

- TWO MATERIALS can become ELECTRICALLY CHARGED when they are RUBBED AGAINST EACH OTHER.
- The materials have become charged with **STATIC ELECTRICITY** meaning ...
- ... that the electricity stays on the material and doesn't move.

You can 'generate' static electricity by rubbing a balloon against a jumper.

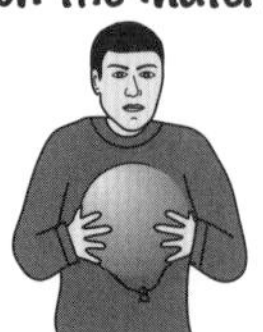

The electrically charged balloon will then attract very small objects.

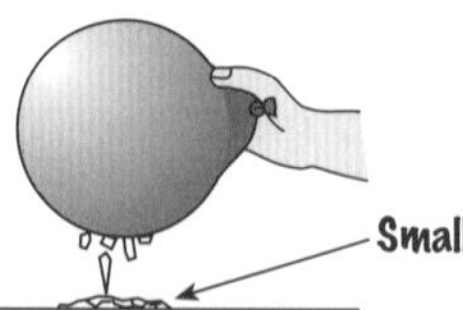

WHY MATERIALS BECOME ELECTRICALLY CHARGED

If two insulating materials are rubbed against each other and become CHARGED then ELECTRONS are 'scraped off' one material ...
... and put on the other material leaving ONE of them **NEGATIVELY CHARGED** and the OTHER **EQUALLY POSITIVELY CHARGED.**

POLYTHENE ROD RUBBED WITH A CLOTH

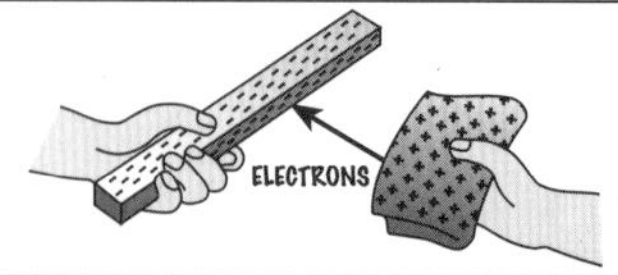

Polythene GAINS electrons ... ... NEGATIVELY CHARGED.

Cloth LOSES electrons ... ... POSITIVELY CHARGED.

PERSPEX ROD RUBBED WITH A CLOTH

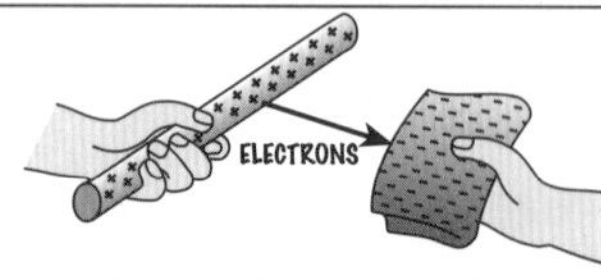

Perspex LOSES electrons ... ... POSITIVELY CHARGED.

Cloth GAINS electrons ... ... NEGATIVELY CHARGED.

REPULSION AND ATTRACTION BETWEEN CHARGED MATERIALS

Very simply ... TWO MATERIALS WITH THE SAME CHARGE ON THEM, REPEL EACH OTHER while ...
... TWO MATERIALS WITH DIFFERENT CHARGE ON THEM, ATTRACT EACH OTHER.

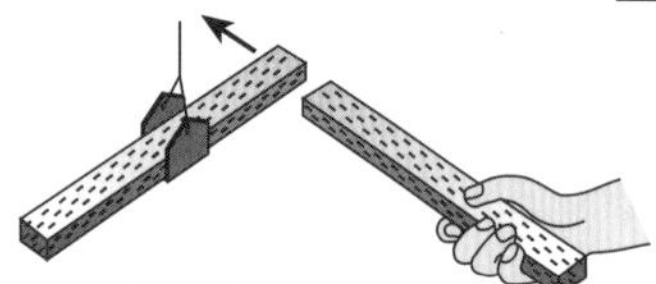

- The SUSPENDED POLYTHENE ROD ...
- ... is REPELLED by the HELD POLYTHENE ROD.

NB. WE WOULD GET THE SAME WITH PERSPEX RODS.

- The SUSPENDED PERSPEX ROD ...
- ... is ATTRACTED by the HELD POLYTHENE ROD.

NB. WE WOULD GET THE SAME IF THE RODS WERE THE OTHER WAY AROUND.

USES OF ELECTROSTATIC CHARGE

ELECTROSTATIC SMOKE PRECIPITATOR

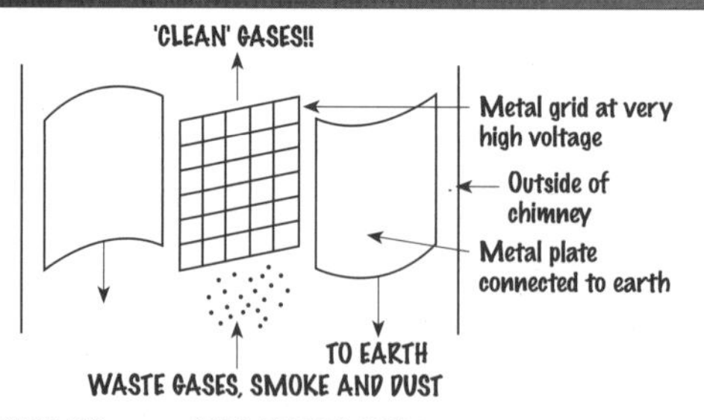

USED AT POWER STATIONS

- SMOKE and DUST PARTICLES are CHARGED UP by ...
- ... VERY HIGH VOLTAGE between GRID and PLATE.
- These charged particles are attracted to the EARTHED METAL PLATES ...
- ... where they then LOSE THEIR CHARGE ...
- ... falling down the chimney where they are then removed.

KEY POINTS:

- The movement of electrons from one material to another creates electrically charged materials.
- Materials with like charge repel each other.
- Materials with unlike charge attract each other.

OBJECTS WITH UNBALANCED CHARGE - EFFECT OF DISCHARGE

- The greater the charge on an insulated object ...
- ... the greater the voltage (potential difference) ...
- ... between the object and earth.
- If there is an earthed conductor near enough, and ...
- ... the voltage gets big enough then a spark may ' jump the gap' ...
- ... as **DISCHARGE** occurs through the ionised air molecules.
- This flow of charge is an electric current.

In this case the electrons run from the object to earth.

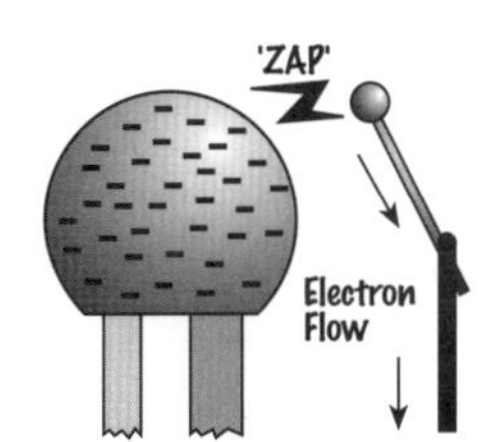

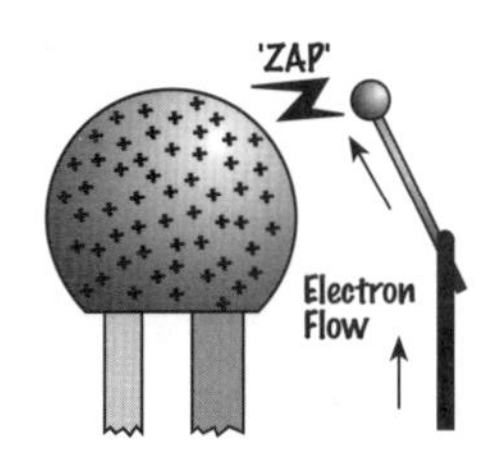

In this case the electrons run from earth to the object.

TWO examples are ...
- LIGHTNING where clouds become charged up ...
- ... by rising hot air until ... DISCHARGE!!

and
- 'STATIC' SPARKS on CLOTHING removed from body.

... while THREE potentially very dangerous examples are:

1. HIGH VOLTAGE CABLES
- Massive charge on cable compared to an earthed object nearby.
- Therefore charge can 'jump' across the gap in suitable conditions.
- Contact is usually FATAL with SEVERE BURNS.
- Casualty may even be thrown from point of contact.

2. ELECTRIC SHOCK
- Severity depends on AMOUNT of ELECTRON FLOW through BODY.
- HEART and MUSCLES are affected causing ...
- ... possible 'MUSCLE SPASM' or DEATH.

3. IGNITION OF FLAMMABLE VAPOURS e.g. Refuelling: (Aircraft)
- Fuel picks up electrons from fuel pipe.
- Fuel tank becomes -ve; fuel pipe becomes +ve.
- This difference can cause a SPARK! ... Possible explosion!!

SOLUTION
- Earth the fuel tank or connect tanker and plane via a conductor.
- EITHER WAY SAFE DISCHARGE WILL OCCUR.

N.B. **GREATER THE CHARGE, THE GREATER THE 'SPARK', GREATER THE FLOW OF ELECTRONS, GREATER THE DANGER!**

CHARGE AND CURRENT - THE RELATIONSHIP

Charge is measured in COULOMBS (where 1 coulomb = 6 million, million, million electrons if you really want to know).

CHARGE (Coulombs) = CURRENT (Amps) x TIME (Secs)

$$Q = I \times t$$

So, if a current of 7 amps flows for 3 secs, 21 coulombs of charge flow.

$$\frac{Q}{I \times t}$$

YOU MUST KNOW THIS FORMULA - SO LEARN IT.

KEY POINTS:

- A flow of charge is an electric current
- Charge (C) = Current (A) x Time (s)

ENERGY IN CIRCUITS I

An **ELECTRIC CURRENT** will flow through an **ELECTRICAL COMPONENT** if there is a **VOLTAGE** or **POTENTIAL DIFFERENCE** (p.d.) across the ends of the component. The amount of current depends on two things:

1. THE POTENTIAL DIFFERENCE (p.d.) ACROSS THE COMPONENT

The **BIGGER the POTENTIAL DIFFERENCE or VOLTAGE** across a component ...
... the **BIGGER the CURRENT** that flows through the component.

(a)

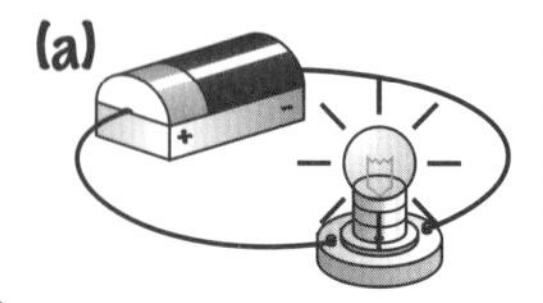

Cell provides p.d. ...
... across the lamp.
A current flows and ...
... the lamp lights up.

(b)

Two cells together provide ...
... a bigger p.d. across the lamp.
A bigger current now flows ...
... and the lamp lights up more brightly.

2. THE RESISTANCE OF THE COMPONENT

COMPONENTS RESIST the FLOW of CURRENT THROUGH THEM. They have **RESISTANCE.**
The **BIGGER the RESISTANCE** of a COMPONENT or COMPONENTS ...
... the **SMALLER the CURRENT** that flows for a **PARTICULAR VOLTAGE.**

(c)

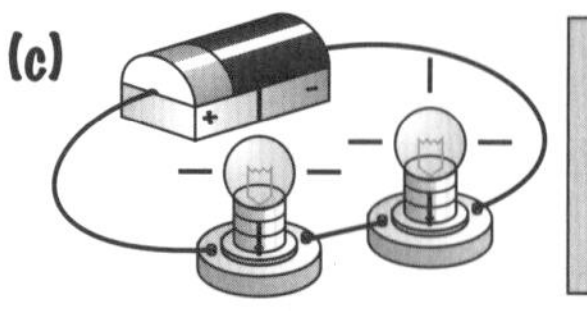

Two lamps have a bigger resistance resulting in a smaller current and a dimmer lamp.

(d)

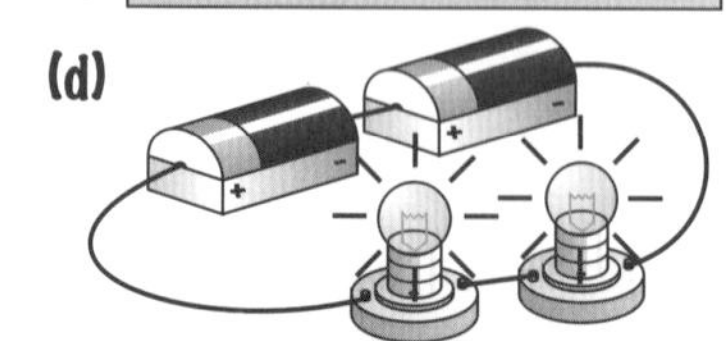

Two cells provide a bigger voltage, resulting in the same current as (a) and a brighter lamp than (c).

... or the **BIGGER the VOLTAGE** needed to maintain a **PARTICULAR CURRENT.**

MEASUREMENT OF POTENTIAL DIFFERENCE AND CURRENT

Here is a very simple circuit ...

AN AMMETER (–A–) ...
... measures the current flowing through the circuit in AMPS. It is always connected in SERIES (see next page).

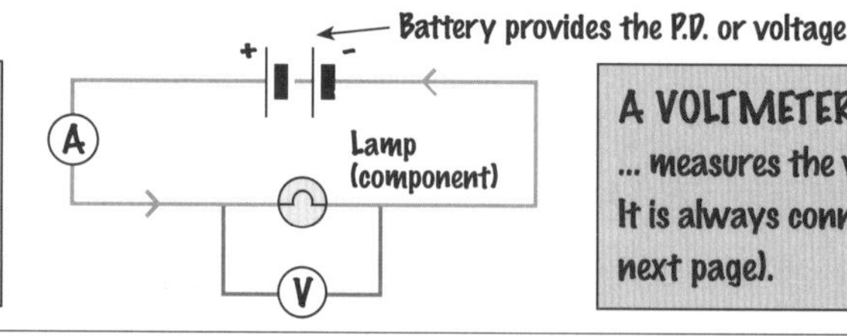

A VOLTMETER (–V–) ...
... measures the voltage across the lamp. It is always connected in PARALLEL (see next page).

ELECTRICAL SYMBOLS

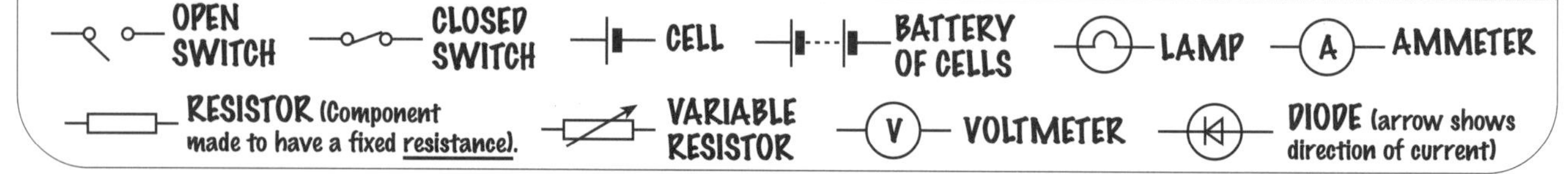

HEATING EFFECT IN RESISTORS

Very simply ...

- MOVING ELECTRONS (CHARGE) collide with ATOMS within a RESISTOR ...
- ... GIVING UP THEIR ENERGY which results in the TEMPERATURE of the RESISTOR INCREASING. Three common electrical appliances using this effect are:

1. Hairdrier — Heating Element (the resistor)

2. Immersion heater — Immersion Heater (the resistor)

3. Light bulb — Tungsten Filament (the resistor)

KEY POINTS:

- The amount of current flowing through a component depends on the Potential difference or voltage across the component and the Resistance of the component.

SERIES CIRCUITS

1. The SAME CURRENT flows through ...
   ... EVERY COMPONENT in the circuit.

$$A_1 = A_2 = A_3$$

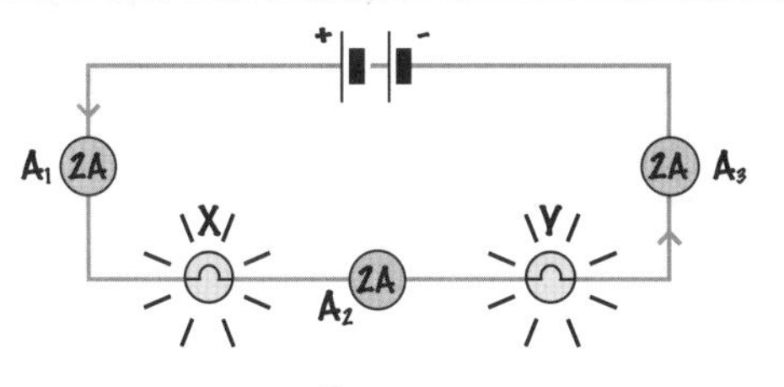

2. The TOTAL VOLTAGE is DIVIDED ...
   ... BETWEEN ALL OF THE COMPONENTS in the circuit.

$$V_{TOTAL} = V_X + V_Y$$

In this case the two bulbs have identical resistances and therefore the voltage is split equally but the voltage could be split 4 and 2 or 5 and 1 (for example) if bulbs of different resistance were used.

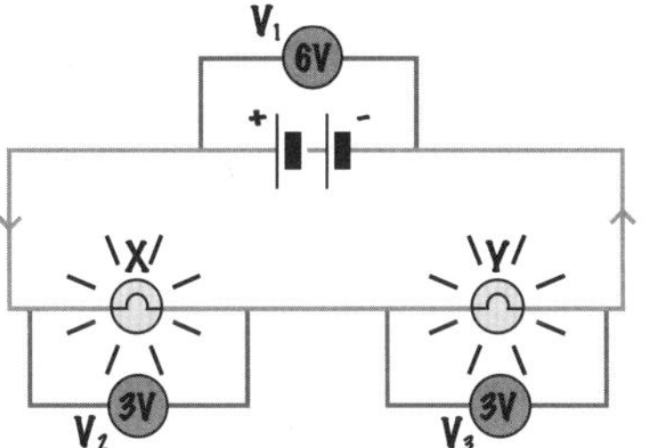

ADVANTAGES
- Supply voltage is divided between the bulbs so it's less dangerous if it is the mains supply of 240V!!
- Less 'drain' on the supply voltage as current supplied is LESS than in a parallel circuit.

DISADVANTAGES
- Can't have one bulb lit on its own, ALL ON or ALL OFF!
- If one bulb fails, they ALL fail.

PARALLEL CIRCUITS

1. The TOTAL CURRENT in the MAIN CIRCUIT ...
   ... is the SUM of the CURRENTS THROUGH EACH COMPONENT.

$$A_1 = A_2 + A_3 = A_4$$

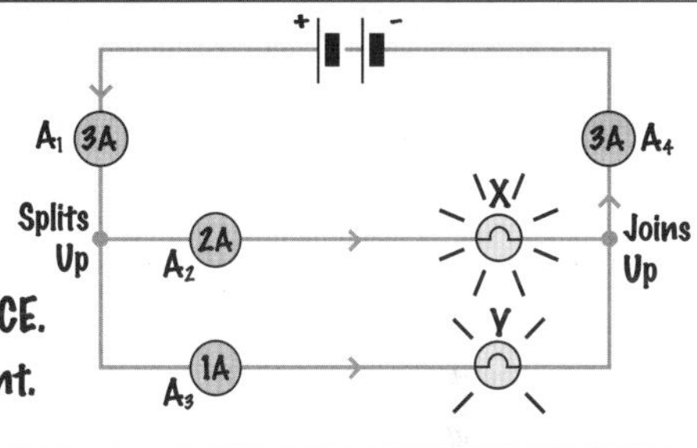

2. The CURRENT through each component depends on its RESISTANCE.
   The greater the resistance of a component, the smaller the current.

Resistance of X is LESS THAN the resistance of Y so the ... current through X is MORE THAN the current through Y!!

3. There is the SAME VOLTAGE ...
   ... ACROSS EACH COMPONENT in the circuit.

$$V_{(across\ x)} = V_{(across\ y)}$$

<u>For light bulbs in series:</u>

ADVANTAGES
- If one bulb fails, then other bulbs are still working.
- Possible to have one bulb lit on its own.

DISADVANTAGES
- All connected directly to supply voltage so it can be dangerous if it is the mains supply of 240V!!
- More 'drain' on the supply voltage as current supplied is MORE than in a series circuit.

CHARGE and VOLTAGE - the relationship

VOLTAGE is the ENERGY TRANSFERRED PER UNIT CHARGE e.g. a 6 Volt battery will transfer 6 Joules of energy to EVERY coulomb of charge that passes through it.

$$\text{VOLTAGE (Volts)} = \frac{\text{ENERGY (Joules)}}{\text{CHARGE (Coulombs)}}$$

E
V × Q

YOU MUST KNOW THIS FORMULA - SO LEARN IT.

KEY POINTS:

- Components in series have the same current flowing through them but the total voltage is divided up
- Components in parallel have the same voltage across them but the total current is divided up
- Voltage (V) = Energy (J) ÷ Charge (C)

ENERGY IN CIRCUITS III - Resistance

The CURRENT in an electrical circuit will CHANGE if there is a ...

- ... CHANGE in the VOLTAGE within the circuit and/or a ...
- ... CHANGE in the RESISTANCE within the circuit ...

... where resistance is a measure of how hard it is to get the current through a conductor or component at a particular voltage. It's unit is OHMS (Ω).

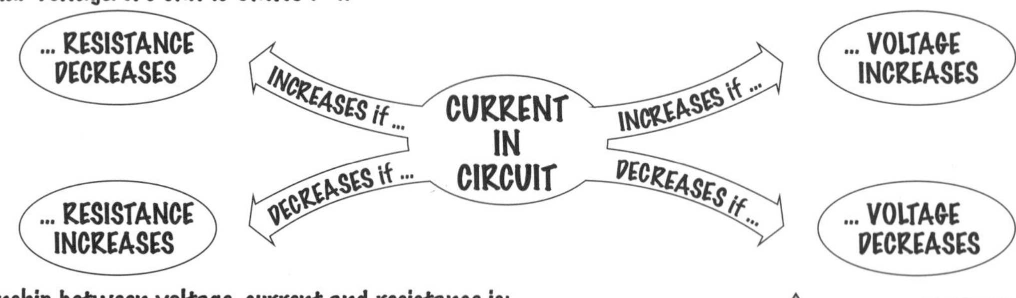

The relationship between voltage, current and resistance is:

VOLTAGE (Volts) = CURRENT (Amps) x RESISTANCE (Ohms)

$V = {}^{*}I \times R$

* (we used the letter I for current)

EXAMPLE

A 12 volt battery is connected to a single lamp and a current of 3 Amps passes through the lamp. What is the resistance of the lamp?

Using our equation, Resistance $= \frac{\text{Voltage}}{\text{Current}}$, $R = \frac{12V}{3A}$, R = 4 Ohms (Ω)
(which is rearranged)

CURRENT - VOLTAGE GRAPHS

These explore the variation of current with voltage through a component.

① A RESISTOR MAINTAINED AT CONSTANT TEMP.

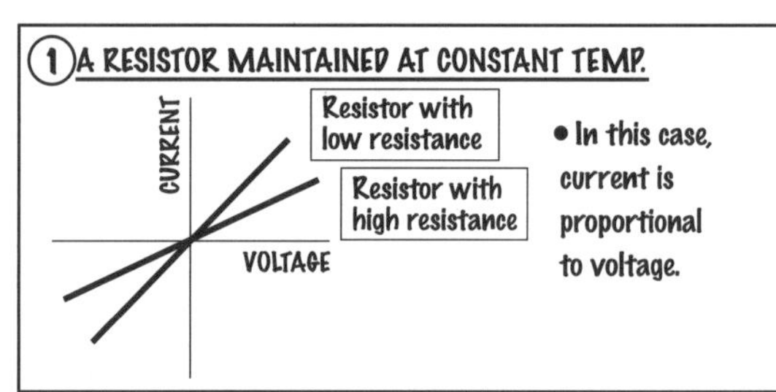

- In this case, current is proportional to voltage.

② WIRES MADE OF DIFFERENT METALS AT CONST. TEMP.

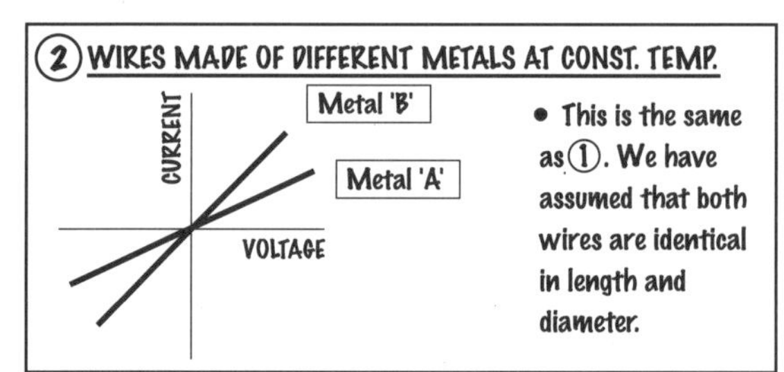

- This is the same as ①. We have assumed that both wires are identical in length and diameter.

③ A FILAMENT BULB

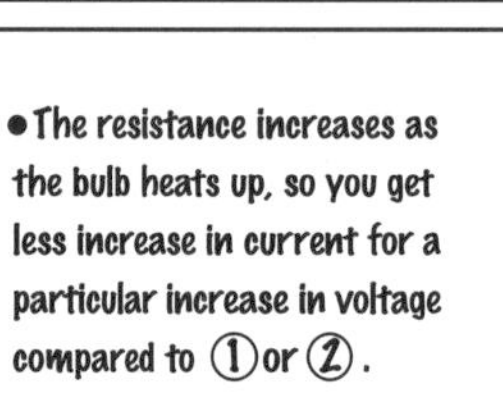

- The resistance increases as the bulb heats up, so you get less increase in current for a particular increase in voltage compared to ① or ②.

④ A DIODE

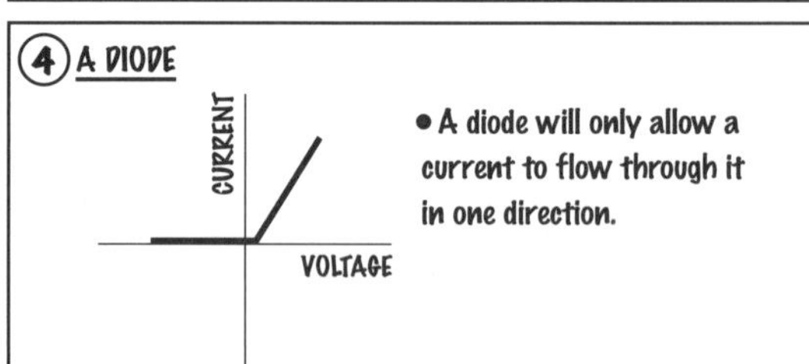

- A diode will only allow a current to flow through it in one direction.

⑤ LIGHT DEPENDENT RESISTORS (L.D.R)

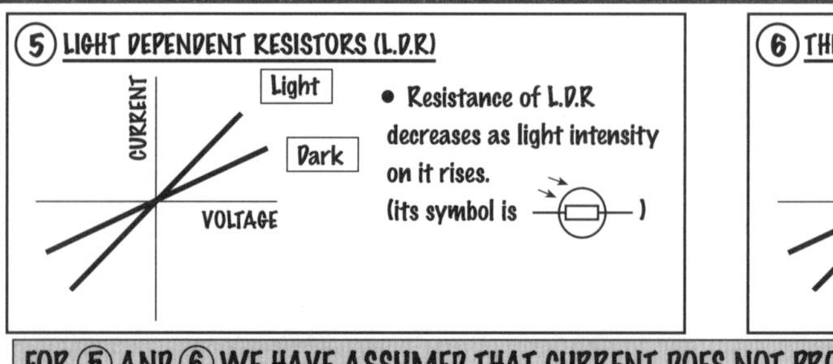

- Resistance of L.D.R decreases as light intensity on it rises. (its symbol is)

⑥ THERMISTOR

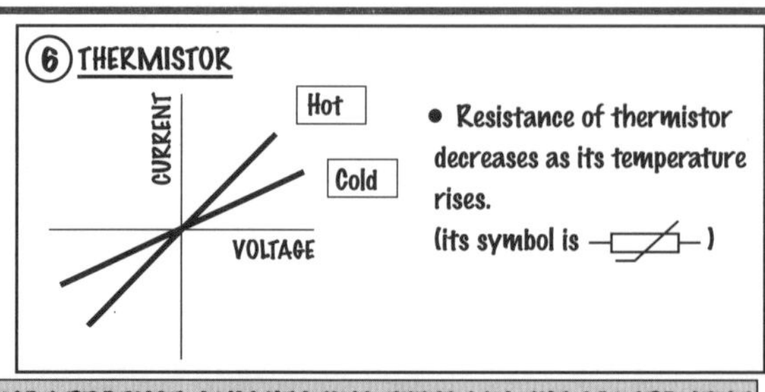

- Resistance of thermistor decreases as its temperature rises. (its symbol is)

FOR ⑤ AND ⑥ WE HAVE ASSUMED THAT CURRENT DOES NOT PRODUCE ANY HEATING EFFECT IN THE COMPONENT

KEY POINTS:

- Voltage (V) = Current (A) x Resistance (Ω)
- Current-Voltage graphs can be used to describe the resistance of a component.

POWER AND MAINS ELECTRICITY I - Fuses

POWER

When an ELECTRIC CURRENT flows through a circuit ...

- ... ENERGY is TRANSFERRED from the BATTERY or SUPPLY VOLTAGE ...
- ... to the COMPONENTS in that circuit.

The **RATE** of this **ENERGY TRANSFER** is the **POWER** in Joules/sec or **WATTS**

The relationship: **ELECTRICAL POWER (Watts) = VOLTAGE (Volts) x CURRENT (Amps)**

FUSES

Very simply ...

- A FUSE is a SHORT, THIN piece of WIRE ...
- ... with a LOW MELTING POINT.
- When the CURRENT passing through it EXCEEDS ...
- ... the CURRENT RATING of the fuse, ...
- ... the fuse wire gets HOT and BURNS OUT or BREAKS.
- This PREVENTS DAMAGE to CABLE or APPLIANCE through the possibility of OVERHEATING.

TOO LARGE A CURRENT → GREATER THAN CURRENT RATING OF FUSE → FUSE BURNS OUT → CIRCUIT IS BROKEN → NO CURRENT FLOWS → CABLE OR APPLIANCE IS PROTECTED

EXAMPLES

1. A domestic iron has a rating plate stuck on it as shown below. Calculate the current that passes through the iron when in use and the current rating of the fuse which should be fitted into its plug if a 3A, 5A, 10A and 13A fuse are available.

POWER → 900W   VOLTAGE → 240v-50Hz
WELLMAN
SUPERSTEAM
SERIAL No 6161623PW

DOMESTIC IRON RATING PLATE

THE RATING PLATE GIVES US THE POWER AND VOLTAGE and so ...

Using our equation above: $\text{Current} = \frac{\text{Power}}{\text{Voltage}}$, $I = \frac{900W}{240V}$, $I = 3.75$ Amps
(which is rearranged)

Fuse chosen for plug should be 5A because its just above the normal current flowing while ...

- A 3A fuse if chosen would burn out immediately (fairly obvious) and ...
- ... a 10A or 13A fuse would work BUT both would allow a large and potentially dangerous current to flow before burn out occurred.

2. A tablelamp connected to the mains voltage at 240V contains a filament of resistance 960Ω.
What current rating should the fuse in its plug have?
THIS TIME WE ARE GIVEN THE VOLTAGE AND RESISTANCE and so ...
Using the equation from E.M.5: $\text{Current} = \frac{\text{Voltage}}{\text{Resistance}}$, $I = \frac{240V}{960}$, $I = 0.25A$
(which is rearranged)
Fuse chosen for plug should be 3A.

KEY POINTS:

- Electrical Power (W) = Voltage (V) x Current (A)
- A fuse is a safety device used to protect a cable or an appliance.

3 PIN FUSED PLUG

N.B. No bare wires showing around screws!!

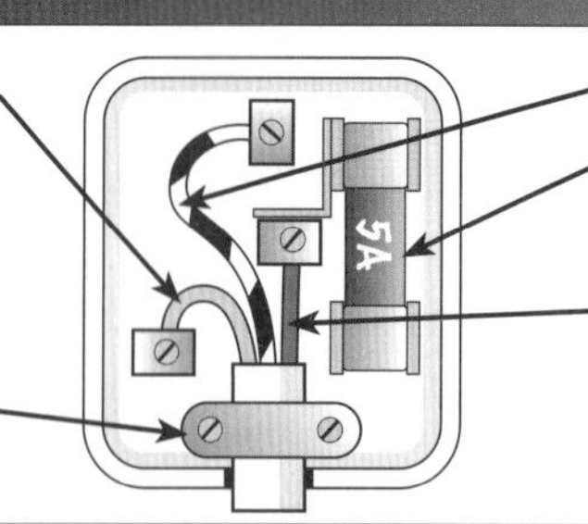

NEUTRAL WIRE (Blue)
Carries current away from appliance. Stays at a voltage close to zero with respect to earth.

CABLE GRIP
Must be tight to stop the cable moving.

EARTH WIRE (Green = Yellow)

FUSE
Always part of the live wire.

LIVE WIRE (Brown)
Carries current to appliance. Alternates between a positive and negative voltage with respect to neutral terminal.

FUSES AND EARTHING

- A fuse is a short thin piece of wire with a low melting point.
- These are placed inside the 3-pin plug (see above) ...
- ... or sometimes actually within an appliance.

All appliances with outer metal cases must be earthed. The outer case is connected to the earth pin in the plug through the earth wire ...

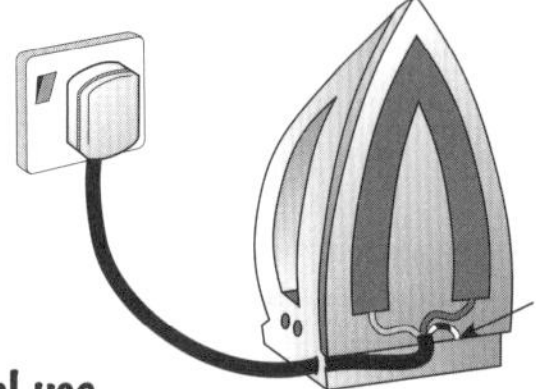

In normal use ...
- ... a current from the live terminal of the mains supply ...
- ... which alternates between a POSITIVE and NEGATIVE VOLTAGE ...
- ... passes through the heating element ...
- ... causing the iron to get hot.

The current then returns through the neutral terminal ...
- ... which is at a VOLTAGE close to zero ...
- ... with respect to Earth.

However ...
- ... if a stray wire from the LIVE touches the metal case ...
- ... the case will become live and ...
- ... there is a surge of current to earth ...
- ... via this 'SHORT CIRCUIT' which causes ...
- ... the fuse wire to melt, ...
- ... which breaks the circuit.

'LIVE' CASING → SHORT CIRCUIT → CURRENT "SURGES" TO EARTH → FUSE MELTS → CIRCUIT BROKEN

INSULATION AND DOUBLE INSULATION

ALL ELECTRICAL APPLIANCES should have proper INSULATION where ...

... within the plug ...
NO FRACTURE OR DAMAGE TO SHEATHING OF WIRES WITH NO BARE WIRE IN VIEW AT THE TERMINALS

... while ...

... between the plug and appliance ...
NO FRACTURE OR DAMAGE TO SHEATHING OF CABLE WHERE BARE WIRES MAY SHOW TO MAKE CONTACT WITH USER.

... while ...

... within the appliance ...
ALL WIRES AND 'LIVE METAL PARTS' SHOULD BE PROPERLY INSULATED WITH NO CONTACT WITH OUTSIDE CASING.

DOUBLE INSULATION

Some appliances are DOUBLE INSULATED where ...
- ... all METAL PARTS INSIDE the appliances are COMPLETELY INSULATED from ...
- ... any OUTSIDE PART of the appliance which MAY BE HANDLED.
- These appliances DO NOT HAVE AN EARTH WIRE although they are still PROTECTED BY A FUSE.

CIRCUIT BREAKERS

Circuit breakers are RESETTABLE FUSES while a ...
... RESIDUAL CIRCUIT BREAKER ...

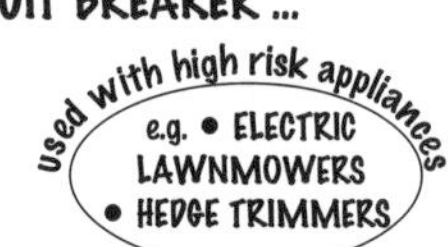

- Detects if there is a DIFFERENCE between ...
- ... the CURRENT in the LIVE and NEUTRAL WIRE (normally the same).
- When this difference is GREATER than a SAFE LEVEL ...
- ... the circuit is BROKEN.

THIS SWITCHES THE CURRENT OFF MUCH FASTER THAN A FUSE.

KEY POINTS:

- A 3 pin plug consists of a live wire, neutral wire, earth wire, fuse and cable grip.
- All appliances with outer metal cases are earthed.

MAINS ELECTRICITY III - Paying for Electricity EM 8

THE ELECTRICITY METER IN YOUR HOME

Your meter at home measures the number of Kilowatt-hours of electricity you use. These are sometimes referred to simply as UNITS of electricity, and are used because Joules are too small a unit of energy.

THE KILOWATT - HOUR

The Kilowatt-hour is a unit of ENERGY ...
... please remember it is NOT a unit of power- that's the kilowatt!!
A electrical appliance transfers 1kWh of energy if it transfers energy at the rate of 1 kilowatt for one hour.

A 200 watt T.V. set  ... transfers 1kWh of energy if it is switched on for 5 hours.

A 500 watt vacuum cleaner  ... transfers 1kWh of energy if it is switched on for 2 hours.

A 1,000 watt electric fire  ... transfers 1kWh of energy if it is switched on for 1 hour.

CALCULATING THE COST

- How much electrical energy or UNITS an appliance uses, depends on ...
  - HOW LONG THE APPLIANCE IS SWITCHED ON
  - HOW FAST THE APPLIANCE TRANSFERS ENERGY (ITS POWER!)

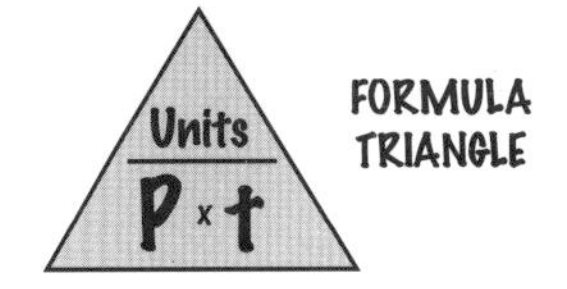

This is our relationship:-

UNITS (kilowatt-hours) = POWER (kilowatts) x TIME (hours) Note the units!

However, to calculate the cost of the electricity 'used', we need to know ...

- HOW MUCH ENERGY HAS BEEN TRANSFERRED i.e. NO OF UNITS USED UP (Remember a 'UNIT' is a kWh).
- HOW MUCH EACH UNIT COSTS.

BOTH THESE FORMULAS ARE GIVEN

This is our relationship:-

COST OF ELECTRICITY SUPPLIED (p) = ENERGY TRANSFERRED (kWh) x PRICE PER UNIT (p/kWh)

KILOWATT - HOUR CALCULATIONS

Remember, we need to use the information above ...

EXAMPLE A 1500 watt electric fire is switched on for 4 hours. How much does this cost if electricity is 9p per unit?

Using the above equation ... **UNITS (kWh) = Power (kW) x time (hours)**

= 1.5 (kW) x 4 (hours)

kilowatts remember!

= 6 kWh (or units)

But, **Cost of electricity supplied = Energy transferred (kWh) x price per unit (p/kWh)**

Therefore Total Cost = 6 x 9

= 54p

Make sure the POWER is in KILOWATTS, AND make sure that the TIME is in HOURS.

KEY POINTS:

- An electrical appliance transfers 1kWh or 1 Unit of energy if it transfers energy at the rate of 1kW for 1 hour.
- Units (kWh) = Power (kW) x Time (hours).
- Cost of electricity used = Energy transferred (kWh) x Price per unit (p/kWh).

MAGNETS - REPULSION AND ATTRACTION

If a MAGNET is allowed to move freely ...

- ... the end of the magnet which points NORTH is called the NORTH (seeking) POLE ...
- ... and the end which points SOUTH is called the SOUTH (seeking) POLE.

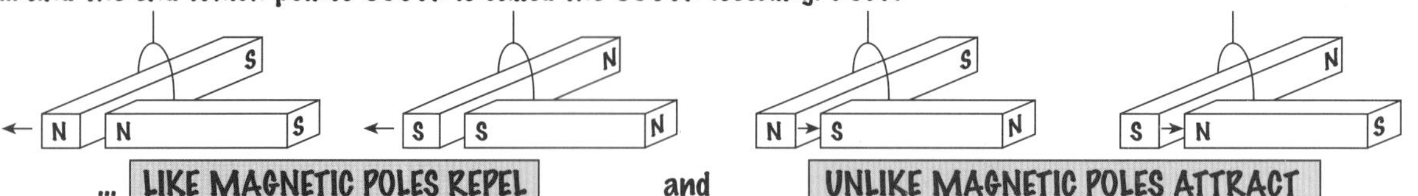

... **LIKE MAGNETIC POLES REPEL** and **UNLIKE MAGNETIC POLES ATTRACT**

MAGNETIC FIELDS

THREE VERY IMPORTANT FACTS:

1. A MAGNETIC FIELD is produced by ...
   - ... a **PERMANENT MAGNET** or ...
   - ... a **CONDUCTOR (WIRE) CARRYING AN ELECTRIC CURRENT**
   - This is an ELECTROMAGNET, a magnet that can be switched ON and OFF.
2. A MAGNETIC FIELD is a space in which ...
   - ... a **PERMANENT MAGNET** or ...
   - ... a **CONDUCTOR (WIRE) CARRYING AN ELECTRIC CURRENT** ...
   - ... EXPERIENCES A FORCE.
3. A MAGNETIC FIELD can be represented by ...
   - ... **LINES OF MAGNETIC FORCE** which show for each point ...
   - ... the DIRECTION OF THE FORCE ...
   - ... which would act on the NORTH (seeking) POLE of another magnet.

Magnetic field patterns

1. BAR MAGNET (permanent field)

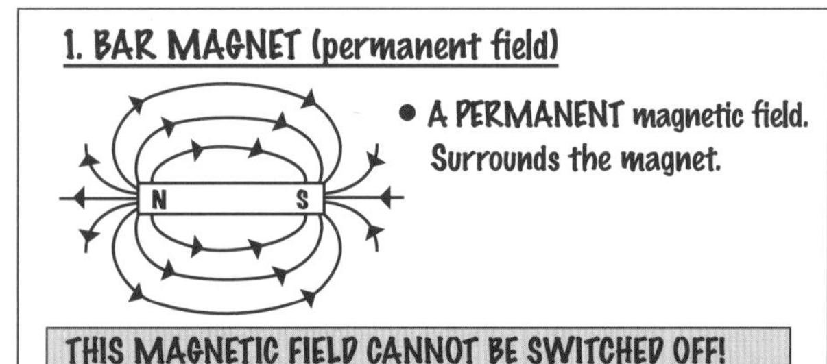

- A PERMANENT magnetic field. Surrounds the magnet.

THIS MAGNETIC FIELD CANNOT BE SWITCHED OFF!

2. A SINGLE WIRE CARRYING A CURRENT (weak electromagnet)

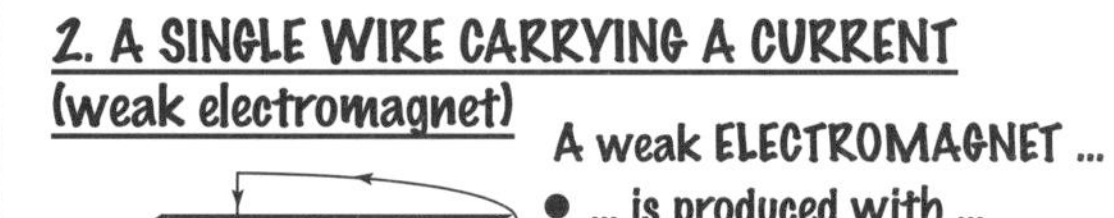

A weak ELECTROMAGNET ...

- ... is produced with ...
- ... circular magnetic field lines

Reversing the current ...

- ... reverses direction of field.

NO CURRENT→NO MAG. FIELD→NO ELECTROMAGNET

3. A SOLENOID (COIL OF WIRE) CARRYING A CURRENT (stronger electromagnet)

- This ELECTROMAGNET has a North and South pole just like an ordinary magnet and you can tell which is which in the following way:

LOOK DIRECTLY AT THE END OF THE COIL YOU ARE CONSIDERING,

- Reversing the current, reverses the POLES of this electromagnet.

NO CURRENT→ NO MAG. FIELD→ NO ELECTROMAGNET

INCREASING THE STRENGTH OF YOUR ELECTROMAGNETS

There are 3 ways:-

- INCREASE THE NUMBER OF TURNS ON THE **COIL**.
- INCREASE THE SIZE OF THE **CURRENT**.
- PLACE A SOFT IRON CORE INSIDE THE **COIL**.

Remember the "three **C**'s", Coil, Current, and Core

KEY POINTS:

- Like magnetic poles repel
- Unlike magnetic poles attract
- The strength of an electromagnet is increased by increasing the number of turns, size of current or by placing a soft iron core inside the coil.

On the previous page we said the electromagnets unlike ordinary (bar) magnets can be switched ON and OFF - very quickly if need be.
This makes them particularly handy for causing USEFUL MOVEMENT in various devices:

TWO CLEVER USES OF ELECTROMAGNETS

1. THE ELECTRIC BELL

Soft Iron Arm
Contact breaks
Electromagnet

- Power switched on causes field ...
- ... which pulls SOFT IRON arm down ringing the bell ...
- ... and breaking the contact causing the arm to return ...
- ... which repeats the whole process.

2. THE RELAY

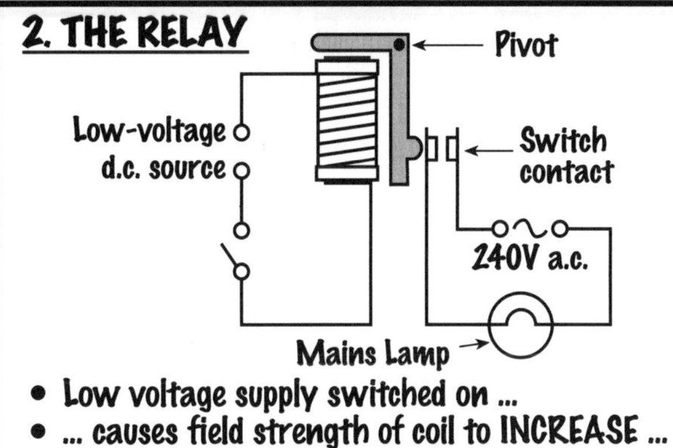

- Low voltage supply switched on ...
- ... causes field strength of coil to INCREASE ...
- ... pulling down a SOFT IRON pivot ...
- ... which causes the 240V supply to be switched on.

These two are mentioned specifically in the syllabus. You MUST be able to explain how they work if you are given the diagram.

Their KEY FEATURES are:
- They all have **COILS** ...
- ... which attract **SOFT IRON** bits when they are switched on ...
- ... causing **ELECTRICAL ENERGY** to convert to **MECHANICAL ENERGY.**

MOTOR PRINCIPLE - SIMPLE DIRECT CURRENT (D.C.) MOTOR

As we have seen when a wire carrying an electric current is placed in a magnetic field it will experience a FORCE. Increasing the FIELD STRENGTH or the CURRENT will increase the size of the force. This is the principle behind the simple D.C. motor:

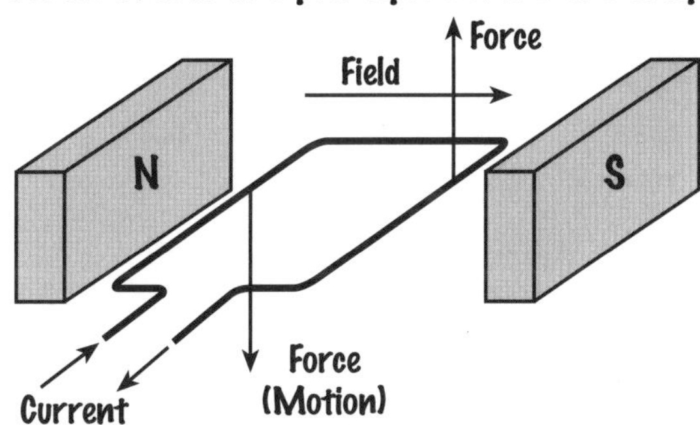

- As current flows through the coil ...
- ... it creates a magnetic field ...
- ... which interacts with the field of the magnet ...
- ... causing a FORCE which ROTATES the COIL.

The direction of the FORCE on each 'side' of the coil can be found using FLEMING'S LEFT HAND RULE, and will change if the direction of current or field changes.

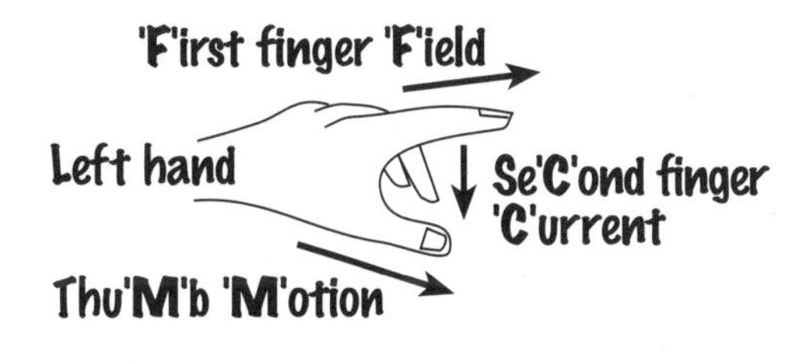

MAKING THE MOTOR TURN FASTER

There are 4 ways:-

- INCREASE THE NUMBER OF TURNS ON THE COIL.
- PLACE A SOFT IRON CORE INSIDE THE COIL.
- INCREASE THE SIZE OF THE CURRENT.
- INCREASE THE MAGNETIC FIELD STRENGTH.

This time, remember three "C's" and an "F" -
COIL, CORE, CURRENT and FIELD STRENGTH.

Yet again you will not be asked to draw the motor but you must be able to explain how it works.

KEY POINTS:

- A wire carrying an electric current placed in a magnetic field will experience a force.
- Increasing the field strength or current will increase the force.

ELECTROMAGNETIC INDUCTION

a.c. and d.c. - comparison as seen on a cathode ray oscilloscope

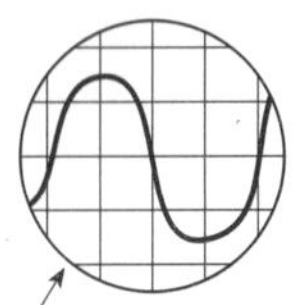

C.R.O. trace

Alternating current (a.c) ....
- Current changes direction of flow ...
- ... back and forth continuously.
- Mains electricity is a.c ...
- ... of frequency 50 Hz ...
- ... i.e. no. of cycles every second.

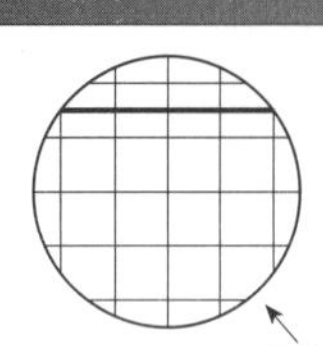

C.R.O. trace

Direct current (d.c) ....
- Current always flows ...
- ... in the same direction.
- Cells and batteries are d.c.

MAKING ELECTRICITY BY ELECTROMAGNETIC INDUCTION

Very simply ...
- If you move a CONDUCTOR (WIRE) or a MAGNET ...
- ... so that the conductor cuts through the lines of force of the magnetic field ...
- ... then a voltage is induced between the ends of the conductor ...
- ... and a current is induced in the conductor if it is part of a complete circuit.

wire (conductor)

"current meter"

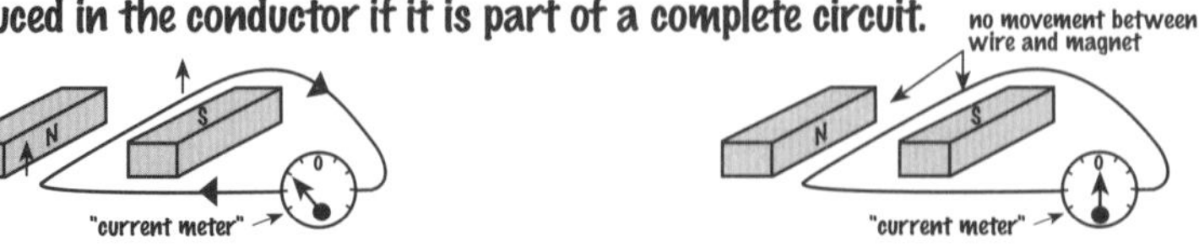

- Moving wire INTO magnetic field ...
- ... induces a current in one direction ...
- ... while moving wire OUT OF mag. field ...
- ... induces current in opposite direction ...
- ... however if there is NO ...
- ... movement of wire or magnet ...
- ... there's no induced current.

THE SAME EFFECT CAN BE SEEN USING A COIL AND A MAGNET ...

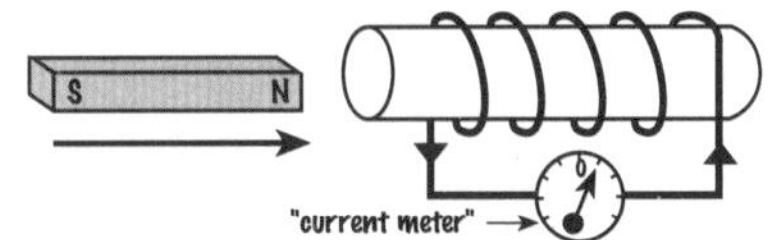

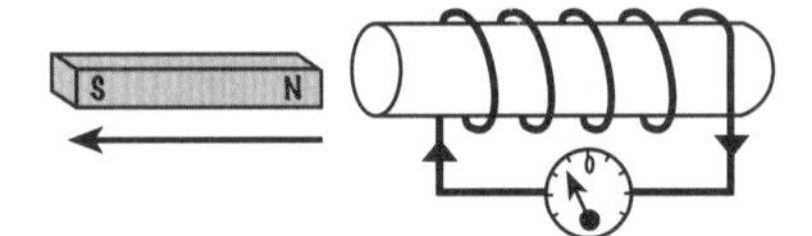

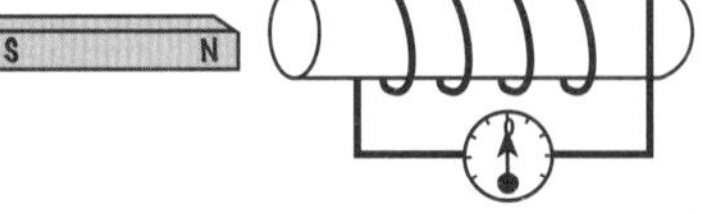

- Moving magnet INTO the coil ...
- ... induces a current in one direction ...
- ... while moving magnet OUT OF the coil ...
- ... induces current in opposite direction ...
- ... however, if there is NO ...
- ... movement of magnet or coil ...
- ... there's no induced current.

The production of an INDUCED VOLTAGE AND CURRENT using a COIL and MAGNET ...
... is called the DYNAMO PRINCIPLE and forms the basis of mains electricity generation.

THE ALTERNATING CURRENT DYNAMO or GENERATOR

... uses the dynamo principle to INDUCE current ...
... which reverses direction each revolution, ...
... as the coil is driven across the lines of magnetic field.

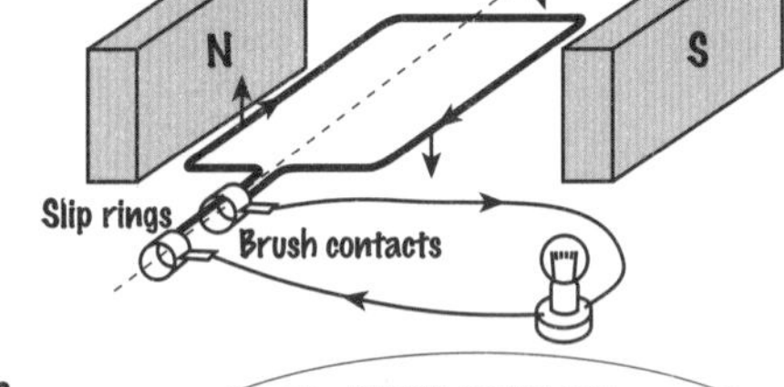

So the general idea for generating electricity is to get the wire crossing the lines of force of the magnetic field as often as possible in order to induce more current. This is logical, is it not?

N.B. The BRUSH CONTACTS are spring-loaded so that they push gently against the SLIP RINGS so that the circuit remains complete. Gradually they wear away and have to be replaced.

This is done by:-

- INCREASING THE SPEED OF MOVEMENT
- INCREASING THE MAGNETIC FIELD STRENGTH
- INCREASING THE NUMBER OF TURNS ON THE COIL
- INCREASING THE AREA OF THE COIL

while INCREASING THESE FOUR THINGS INCREASES THE NUMBER OF LINES OF MAGNETIC FIELD CUT PER SECOND, AND THEREFORE INCREASES THE CURRENT.

(Remember this by "Sloppy Alternator Fails Test"

You will not be asked to draw the a.c dynamo but you must be able to explain its construction and performance.

KEY POINTS:

- A voltage is induced between the ends of a conductor and a current is induced in the conductor if it is part of a complete circuit when it cuts through a magnetic field.

TRANSFORMERS

These are used to: • CHANGE THE VOLTAGE OF AN A.C. SUPPLY.
So that ... • VOLTAGE MAY BE 'STEPPED UP' BEFORE TRANSMISSION.
and then ... • 'STEPPED DOWN' TO WHATEVER VOLTAGE IS NEEDED.

TRANSMISSION OF ELECTRICITY

This is the network of cables which distributes electricity all over the country. BUT ... the HIGHER the CURRENT, the GREATER the AMOUNT OF ENERGY LOST AS HEAT FROM THE CABLES.

So, we need to transmit as low a current as possible but since ...

POWER (watts) = VOLTAGE (volts) x CURRENT (amps) ...

... reducing the current means increasing the voltage (!) to transmit energy at the same rate. THIS IS WHERE TRANSFORMERS COME IN!

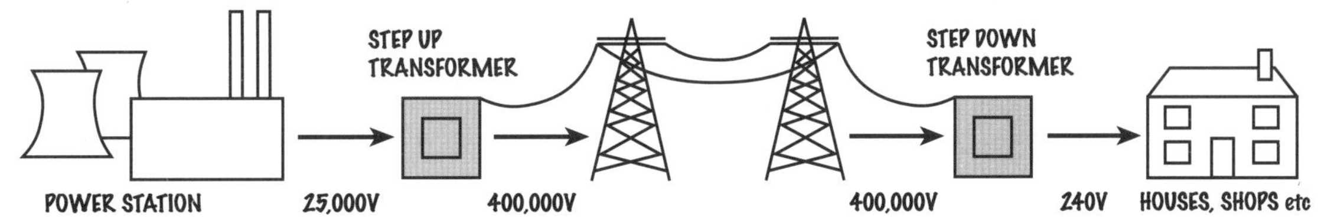

And so for the most effective transmission we need ...

1) **HIGH VOLTAGE** and ...
2) ... **LOW CURRENT** which results in ... 3) **REDUCTION OF HEAT LOSS** in the cables while ...
4) **THICKNESS OF CABLE** is also an important consideration as ideally ...

... THE THICKER THE CABLE → THE LOWER IT'S RESISTANCE → LOWER ENERGY LOSS AS HEAT ...
... however there is a compromise thickness due to the COST AND MASS OF CABLE which ...
... INCREASES with INCREASING THICKNESS.

HOW TRANSFORMERS WORK

PRIMARY COIL (COIL 1)
SECONDARY COIL (COIL 2)
LAMINATED SOFT IRON CORE
50 TURNS
400 TURNS
STEP-UP TRANSFORMER

Transformers consist of ...

- TWO coils called the PRIMARY (coil 1) and SECONDARY (coil 2) ...
- ... wrapped around a LAMINATED SOFT IRON CORE which is made up of ...
- ... sheets of soft iron INSULATED from each other so that it has a HIGH RESISTANCE ...
- ... so that any CURRENT INDUCED WITHIN THE CORE ITSELF IS SMALL ...
- ... resulting in a SMALL HEATING EFFECT.

Transformers work because ...

- ... An ALTERNATING VOLTAGE input in the primary coil (coil 1) ...
- ... causes a continually changing MAGNETIC FIELD ...
- ... which 'cuts across' the secondary coil (coil 2) ...
- ... INDUCING an ALTERNATING VOLTAGE.

THE SIZE OF THE SECONDARY (OR OUTPUT) VOLTAGE DEPENDS ON THE RELATIVE NUMBER OF TURNS ON THE PRIMARY AND SECONDARY COILS:-

$$\frac{\text{VOLTAGE ACROSS COIL 1}}{\text{VOLTAGE ACROSS COIL 2}} = \frac{\text{NUMBER OF TURNS IN COIL 1}}{\text{NUMBER OF TURNS IN COIL 2}}$$

YOU MUST KNOW THIS FORMULA - SO LEARN IT.

Typical exam question: If a voltage of 240v is applied to the PRIMARY COIL in the above diagram, what would the secondary voltage be?

$$\frac{240}{x} = \frac{50}{400} \; ; \; x = \frac{240 \times 400}{50} = \underline{1920 \text{ volts}}$$

OR/ Since there are eight times more coils on the secondary coil, the output voltage will be eight times the input voltage ie, 8 x 240 = 1,920 volts.

KEY POINTS:

• Electricity is transmitted at a high voltage and low current to reduce energy losses. • The size of the output voltage of a transformer depends on the relative number of turns on the primary and secondary coils.

1. What causes static?
2. Name one use for electrostatic charge.
3. What is discharge?
4. Name three very dangerous examples of discharge.
5. What units are used to measure CHARGE?
6. If a current of 21 amps flows for 30 secs, how much charge will flow?
7. What is current?
8. Explain how a) an AMMETER and b) a VOLTMETER must be connected in a circuit.
9. Name three appliances which use the heating effect in resistors.
10. Give one fact about the CURRENT and VOLTAGE in a SERIES CIRCUIT.
11. Give one fact about the CURRENT and VOLTAGE in a PARALLEL CIRCUIT.
12. Give one advantage and one disadvantage of having light bulbs in a SERIES CIRCUIT.
13. Give one advantage and one disadvantage of having light bulbs in a PARALLEL CIRCUIT.
14. What is the relationship between charge and voltage?
15. What is resistance?
16. If the current through a component is 4 amps at a voltage of 8 volts, what is its RESISTANCE? (remember the units)?
17. Draw a current - voltage graph for a) a RESISTOR at constant temperature b) a FILAMENT BULB and c) a DIODE.
18. What is power?
19. What units do we use for POWER?
20. What is a fuse?
21. Work out the fuse needed for an appliance with a power rating of 1kW. Assume 240 volts.
22. Name the three wires in a three pin plug and say what colour they are.
23. What is DOUBLE INSULATION?
24. How does a RESIDUAL CIRCUIT BREAKER work?
25. What is a KILOWATT?
26. What is a KILOWATT-HOUR?
27. Is a Kilowatt-hour a unit of power or energy?
28. A 1000W electric fire is used for 5 hours. What is the cost at 7p per unit?
29. A 2000W electric shower is used for 30 minutes. What is the cost at 7p per unit?
30. Draw the magnetic field pattern for a) a bar magnet b) a single wire carrying a current and c) a solenoid carrying a current.
31. Name three ways in which the STRENGTH of an ELECTROMAGNET may be increased.
32. Explain how electromagnets can be used in two pieces of equipment.
33. What is the MOTOR PRINCIPLE?
34. What do the different digits represent in FLEMING'S LEFT HAND RULE?
35. What is the difference between A.C. and D.C.?
36. Describe how a voltage can be INDUCED across a wire.
37. What do transformers do?
38. Write the equation relating the number of coils on the primary and secondary coils to the primary and secondary voltage.
39. What are the coils wrapped around? Why is this?
40. What is the difference between a 'STEP UP' and a 'STEP DOWN' transformer with regards to turns of wire on the coils?

WAVES

CHARACTERISTICS OF WAVES — W 1

WAVES are
- ... a REGULAR PATTERN OF DISTURBANCE ...
- ... which TRANSFER ENERGY from one point to another ...
- ... without any TRANSFER OF MATTER.

TYPES OF WAVE - shown by using a SLINKY SPRING

1. TRANSVERSE WAVES
- The PATTERN OF DISTURBANCE is ...
- ... at RIGHT ANGLES (90°) to ...
- ... the DIRECTION OF WAVE MOVEMENT.

EXAMPLES
- LIGHT WAVES and all other ...
- ... ELECTRO-MAGNETIC WAVES.
- WATERWAVES.
- WAVES IN ROPES.

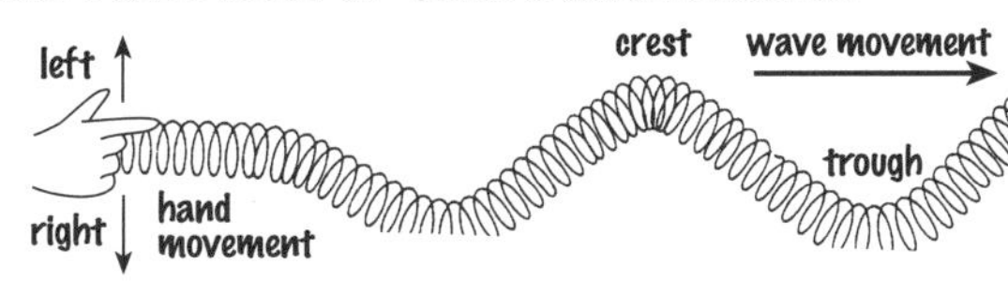

2. LONGITUDINAL WAVES
- The PATTERN OF DISTURBANCE is ...
- ... in the SAME DIRECTION as ...
- ... the DIRECTION of WAVE MOVEMENT.

EXAMPLES
- All SOUND e.g. from simple noises to notes from instruments.

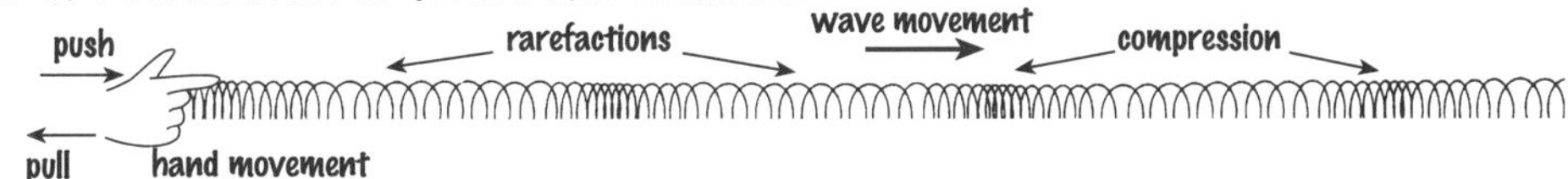

FEATURES OF WAVES - some important definitions

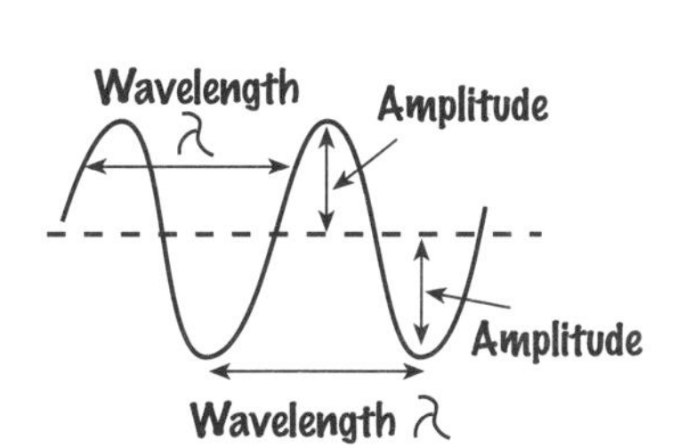

AMPLITUDE (UNITS: cm or m)
- ... is the MAXIMUM DISTURBANCE, ...
- ... i.e. the GREATEST DISPLACEMENT ...
- ... from the NORMAL POSITION

WAVELENGTH (UNITS: m).
- ... is the DISTANCE between ...
- ... CORRESPONDING POINTS on ...
- ... TWO SUCCESSIVE WAVES (symbol = λ)

FREQUENCY (UNITS: Hertz)
- ... is the NUMBER OF WAVES PRODUCED ...
- ... or PASSING A PARTICULAR POINT ...
- ... in ONE SECOND.

THE WAVE EQUATION - applies to ALL waves

For any wave the wavespeed, frequency and wavelength are related by

Wavespeed (m/s) = Frequency (Hz) x Wavelength (m)

YOU NEED TO KNOW THIS FORMULA

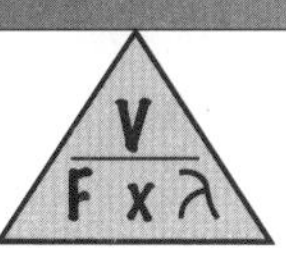

Example $v = f \times \lambda$

1) A sound wave has a frequency of 165Hz and a wavelength of 2m, what is the speed of sound?
Using our equation WAVESPEED = FREQUENCY X WAVELENGTH = 165Hz x 2m = <u>330 m/s</u>

2) Radio 5 live transmits from London on a frequency of 900KHz. If the speed of radio waves is 300,000,000 m/s, what is the wavelength of the waves?
Using our equation, WAVELENGTH = $\frac{\text{WAVESPEED}}{\text{FREQUENCY}} = \frac{300{,}000{,}000 \text{ M/S}}{900{,}000 \text{ Hz}}$ = <u>333.3m</u> ← Must be HERTZ NOT KILOHERTZ!!

KEY POINTS:

- Two types of wave: Transverse and Longitudinal • Amplitude, Wavelength and Frequency are all features of waves • Wavespeed (m/s) = Frequency (Hz) x Wavelength (m).

LIGHT I

VISIBLE LIGHT is a **TRANSVERSE WAVE** and is a small part of the **ELECTROMAGNETIC SPECTRUM** (see W.4).

Light:

- Travels in **STRAIGHT LINES** e.g. formation of **SHADOWS**.
- Travels at a **CONSTANT SPEED** through a **UNIFORM MEDIUM** ... ... the **DENSER** the **MEDIUM**, the **SLOWER** its **SPEED**.
- Can also travel through a **VACUUM**

REFLECTION OF LIGHT

This occurs when light strikes a **SURFACE** or a **MIRROR** resulting in it **CHANGING ITS DIRECTION**.

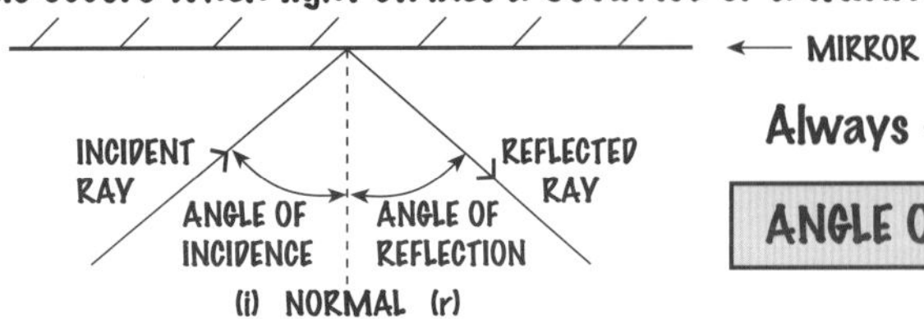

Always -

ANGLE OF INCIDENCE (i) = ANGLE OF REFLECTION (r)

REFRACTION OF LIGHT

- Light **CHANGES DIRECTION**, when it **CROSSES A BOUNDARY** ...
- ... between **ONE TRANSPARENT MEDIUM AND ANOTHER (OF DIFFERENT DENSITY)** ...
- .. unless it meets the boundary along a **NORMAL** (AT 90°)

GLASS OR PERSPEX
AIR

- Rays of light are **REFRACTED** ..
- ... **TOWARDS** the **NORMAL** because ...
- ... they **SLOW DOWN** as ...
- ... they pass from a **LESS DENSE** ...
- ... to a **MORE DENSE MEDIUM**.

- Rays of light are **REFRACTED** ...
- ... **AWAY** from the **NORMAL** because ...
- ... they **SPEED UP** as ...
- ... they pass from a **MORE DENSE** ...
- ... to a **LESS DENSE MEDIUM**.

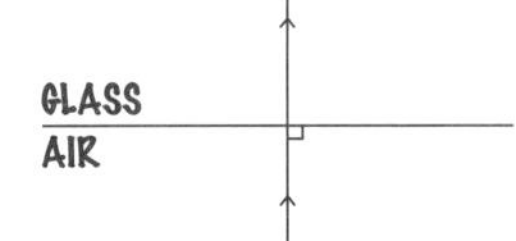

- Rays of light **NOT DEVIATED** since ...
- ... they meet boundary ...
- ... **ALONG THE NORMAL** (at 90°)
- Even so, they still **SLOW DOWN** as ...
- ... they enter the **DENSER MEDIUM**.

REFRACTION OCCURS BECAUSE LIGHT CHANGES SPEED WHEN IT PASSES FROM ONE MEDIUM INTO ANOTHER where ...

- **LESS DENSE to MORE DENSE ... it SLOWS DOWN.**
- **MORE DENSE to LESS DENSE ... it SPEEDS UP.**

DIFFRACTION OF LIGHT

- When **WAVES MOVE THROUGH A GAP** or **PASS AN OBSTACLE** ...
- ... they **SPREAD OUT FROM THEIR EDGES**.

Diffraction is most obvious when:

1.) SIZE OF GAP IS SIMILAR TO WAVELENGTH OF WAVES.

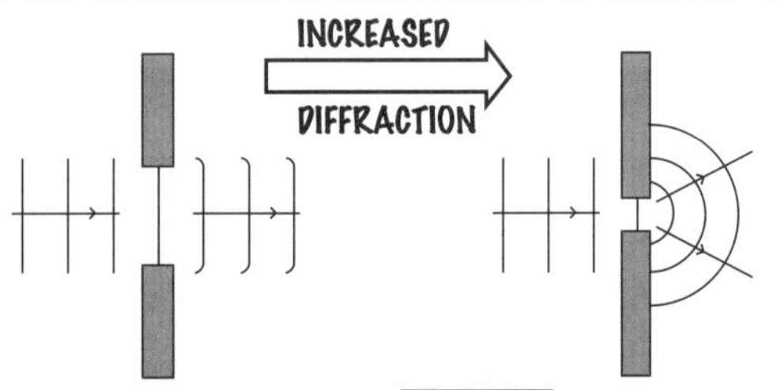

2.) WAVES WHICH PASS OBSTACLES HAVE LONG WAVELENGTHS.

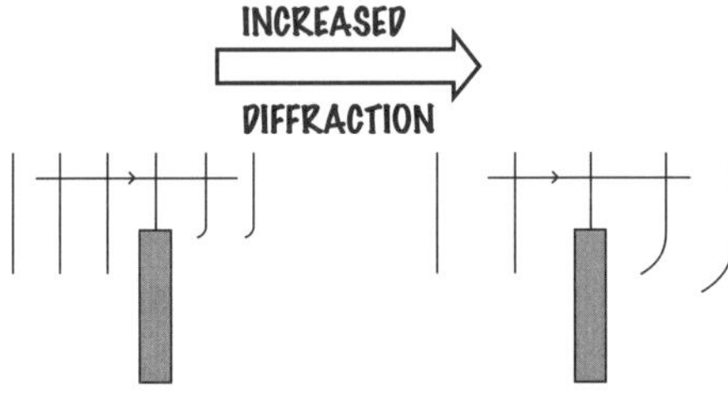

Diffraction also occurs with **SOUND** (see W.5) but because of the longer wavelength, the gaps involved need not be as small as those needed for diffraction of light.

KEY POINTS:

- Light is a transverse wave
- Light can be reflected, refracted and diffracted.

TOTAL INTERNAL REFLECTION - where refraction becomes reflection

Can best be summarised in THREE stages

1) ANGLE OF INCIDENCE < CRITICAL ANGLE

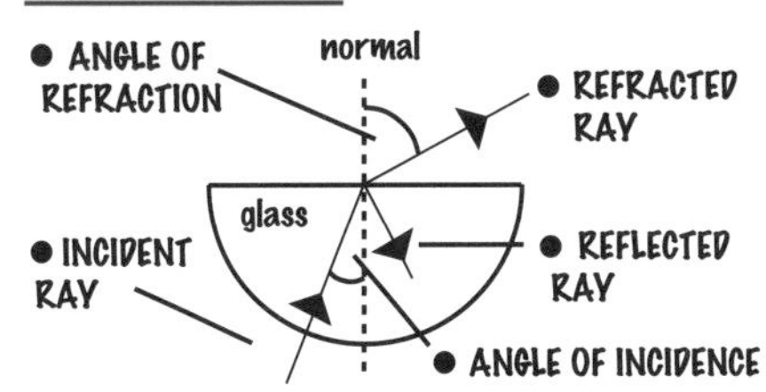

- MOST LIGHT IS REFRACTED but ...
- ... a little is INTERNALLY REFLECTED.

2) ANGLE OF INCIDENCE = CRITICAL ANGLE

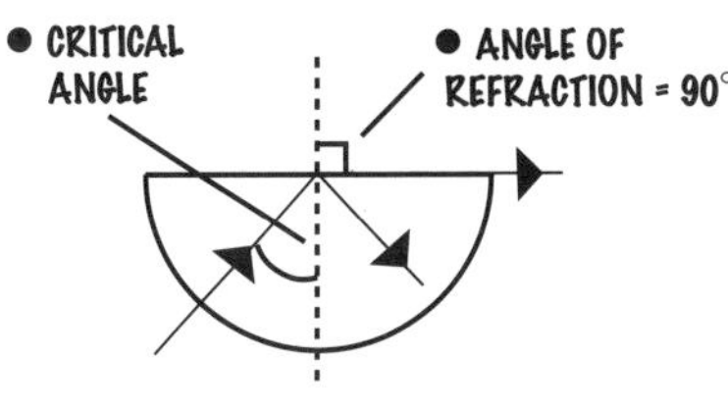

- REFRACTED RAY 'GRAZES' THE BOUNDARY
- ANGLE OF REFRACTION = 90°, AND ...
- ... there is more INTERNAL REFLECTION.

3) ANGLE OF INCIDENCE > CRITICAL ANGLE

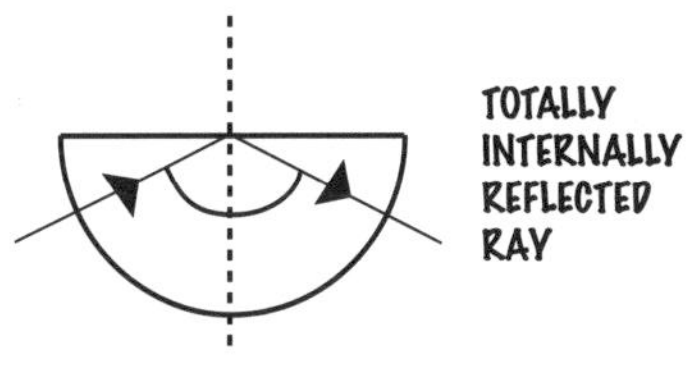

- NO LIGHT IS REFRACTED
- ALL LIGHT IS REFLECTED.
- TOTAL INTERNAL REFLECTION.

USE OF TOTAL INTERNAL REFLECTION

1. OPTICAL FIBRES

ALL ANGLES OF INCIDENCE INSIDE ROD GREATER THAN CRITICAL ANGLE

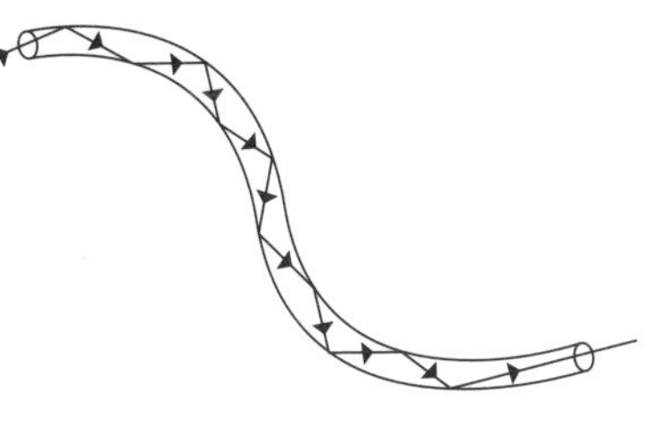

- LONG, FLEXIBLE, TRANSPARENT ROD of ...
- ... very SMALL DIAMETER allows ...
- ... LIGHT TO BE TOTALLY INTERNALLY REFLECTED ...
- ... along its LENGTH.

Examples -

1) ENDOSCOPE - internal viewing of human body.

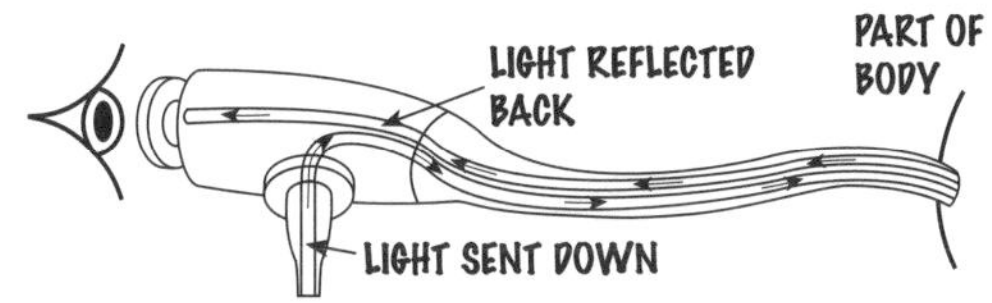

- Consists of BUNDLES OF FIBRES, half of which ...
- ... TRANSMIT LIGHT TO THE PART BEING VIEWED ...
- ... while HALF TRANSMIT REFLECTED LIGHT BACK.

2) TELECOMMUNICATIONS.

- Information can be TRANSMITTED via PULSES of light travelling along OPTICAL FIBRES.
- GREATER CAPACITY than sending ELECTRICAL SIGNALS through CABLES of SAME DIAMETER ...
- ... with LESS WEAKENING of SIGNAL along the way.

2. REFLECTING PRISMS

Can be used to REFLECT LIGHT through ...

a) ... 90°

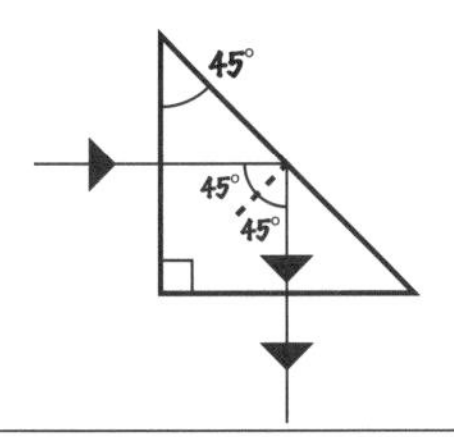

- ALL ANGLES OF INCIDENCE ...
- ... INSIDE PRISM ARE 45° ...
- ... i.e. GREATER THAN CRITICAL ANGLE.

b) ... 180°

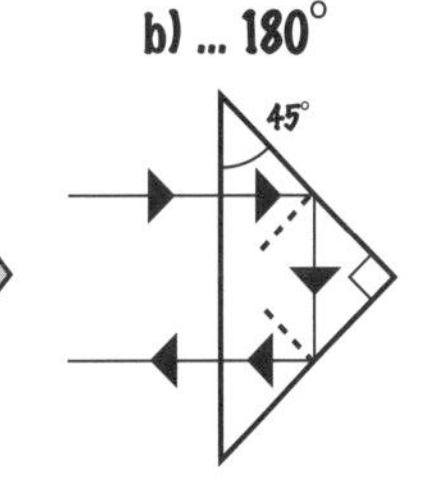

PERISCOPE - For looking over the top of tall objects.

- TWO PRISMS needed as shown.
- ALL LIGHT is REFLECTED ...
- ... better than a MIRROR PERISCOPE ...
- ... as some light is absorbed.

PRISM BINOCULARS

- Contains TWO PRISMS whose job is ...
- ... produce a FINAL UPRIGHT IMAGE ...
- ... for the user.

BICYCLE REFLECTORS, CATS EYES

- Used on roads.

KEY POINTS:

- Total Internal Reflection occurs when the angle of incidence is greater than the Critical Angle.

THE ELECTROMAGNETIC SPECTRUM

Some facts-
- ALL TRANSVERSE WAVES OF DIFFERENT WAVELENGTH AND FREQUENCY ...
- TRAVEL AT THE SAME SPEED (300,000,000 m/s) THROUGH THE AIR ...
- ... OR THROUGH A VACUUM.

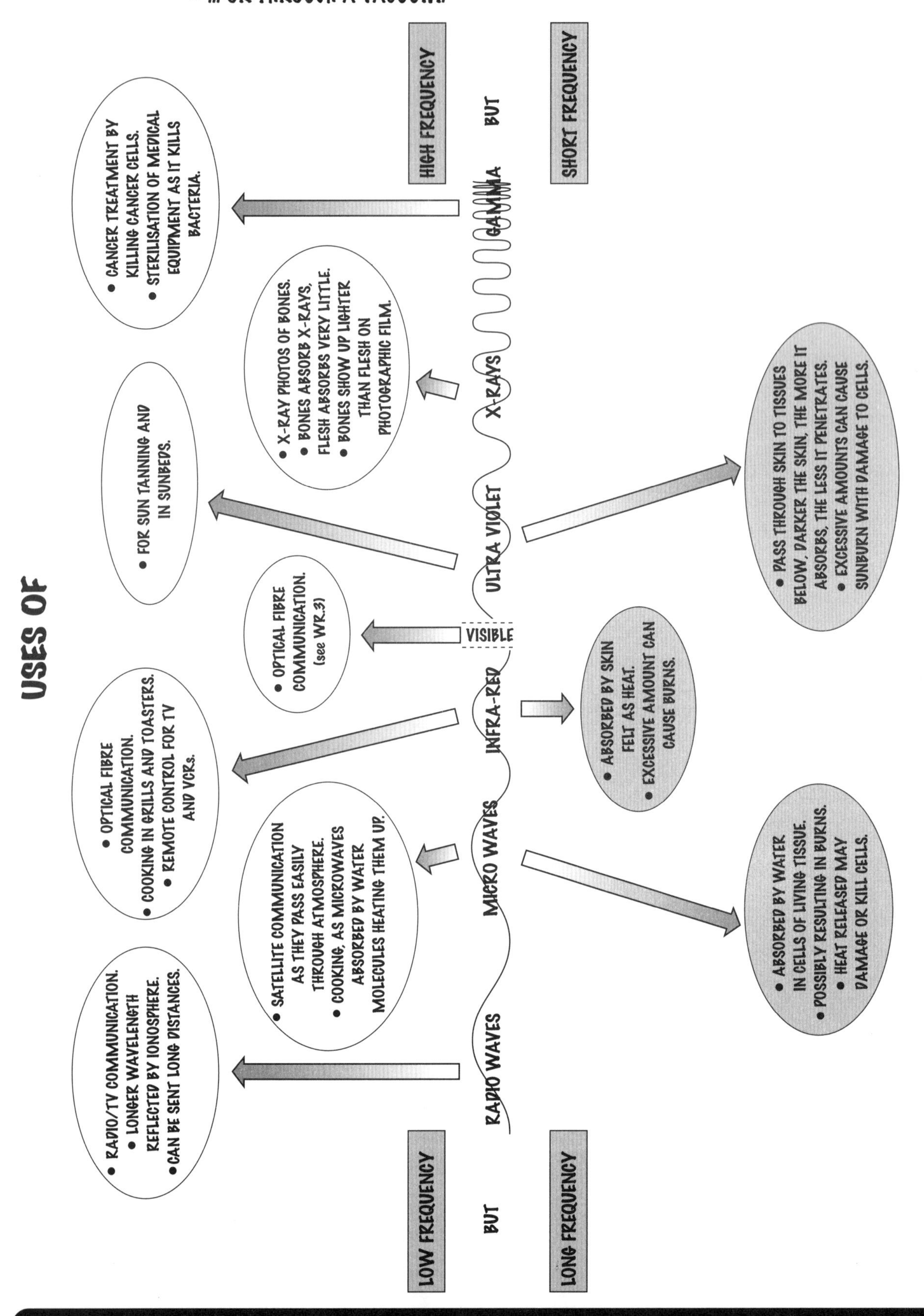

KEY POINTS:

- Radiowaves, Microwaves, Infra-red, Visible, Ultra Violet, X-rays and Gamma rays make up the E-M spectrum.
- They are all transverse waves of different wavelength and frequency.

SOUND I

Sound is produced when something VIBRATES backwards and forwards. This results in the VIBRATION OF A MEDIUM which transfers the sound along it as LONGITUDINAL WAVES (see W.1)

Sound:

- Travels at a CONSTANT SPEED through a UNIFORM MEDIUM ...
- ... the DENSER the MEDIUM, the FASTER its SPEED (opposite of light) although it ...
- can NOT travel through a VACUUM (unlike light).
- is REFLECTED from HARD SURFACES, those reflections are called ECHOES.
- is REFRACTED when it passes into a DIFFERENT MEDIUM.

- is DIFFRACTED by EDGES, so sound can be heard around CORNERS ...
- ... or in the SHADOW of BUILDINGS.

DISPLAYING SOUND WAVES

An instrument called a CATHODE RAY OSCILLOSCOPE (CRO) can be used to display and analyse REGULAR SOUND WAVES.

Using a CRO we can make a comparison of their **LOUDNESS** and **PITCH**.

1. LOUDNESS

- GREATER THE AMPLITUDE (See W.1) ...

... GREATER THE ENERGY CARRIED BY THE WAVE ...

... GREATER THE LOUDNESS.

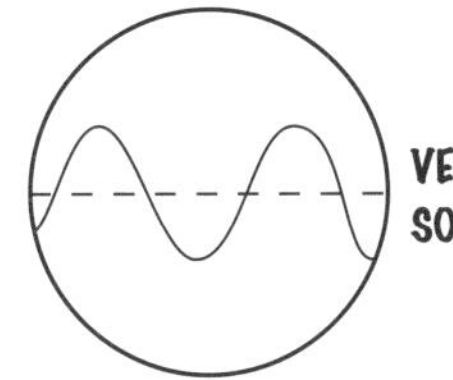

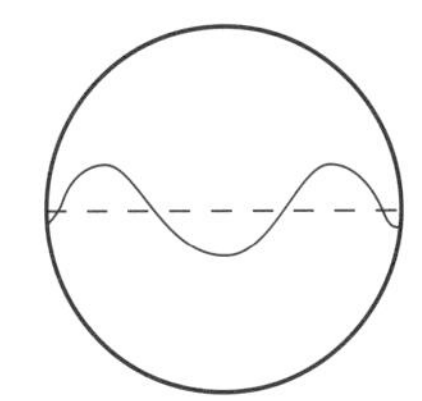

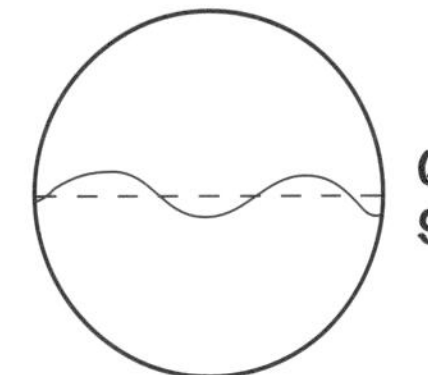

2. PITCH

- GREATER THE FEQUENCY (See W.1) ...

... GREATER THE PITCH.

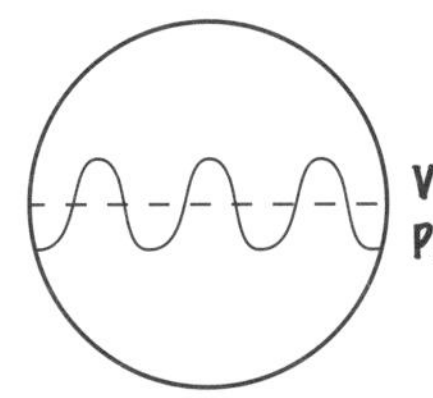

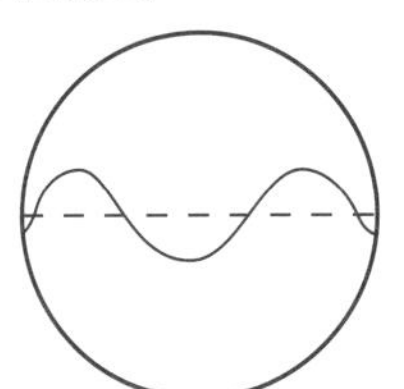

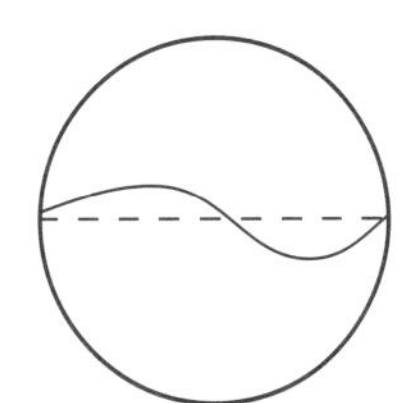

MICROPHONES AND LOUDSPEAKERS

MICROPHONE:

- an energy changer.

- converts SOUND VIBRATIONS ...
- ... into ELECTRICAL VIBRATIONS ...
- ... of the SAME FREQUENCY.

LOUDSPEAKER:

- an energy changer.

- OPPOSITE OF MICROPHONE
- converts ELECTRICAL VIBRATIONS ...
- ... into SOUND VIBRATIONS ...
- ... of the same FREQUENCY.

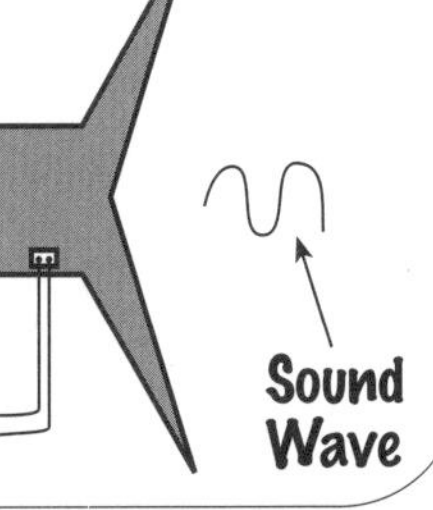

KEY POINTS:

- Sound is a longitudinal wave
- Sound can be reflected, refracted and diffracted.
- Loudness depends on amplitude and pitch depends on frequency.

ULTRASONIC WAVES

These are **SOUND WAVES** of **FREQUENCIES GREATER** than **20,000 Hz** i.e. above the **UPPER LIMIT** of the **HEARING RANGE** for **HUMANS**. They are made by **ELECTRONIC SYSTEMS** which produce **ELECTRICAL OSCILLATIONS** which are passed into a **LOUDSPEAKER**.

Uses of Ultrasonic Waves

1. PRE - NATAL SCANNING (FOETAL IMAGING IN OBSTETRICS)

- Ultrasonic waves sent into **BODY** by **SCANNER** placed in **GOOD CONTACT WITH THE SKIN**.
- **PARTIAL REFLECTIONS** are received by **PROBE** in scanner from any **SURFACES** or **BOUNDARIES** ...
- ... within the body which have a **DIFFERENT DENSITY** or **STRUCTURE**.
- The **TIME DELAY** of the **REFLECTIONS** is a measure of the **DEPTH** of the **REFLECTING SURFACE**.
- The reflected waves are **PROCESSED** to produce a **VISUAL IMAGE** on a **SCREEN**.

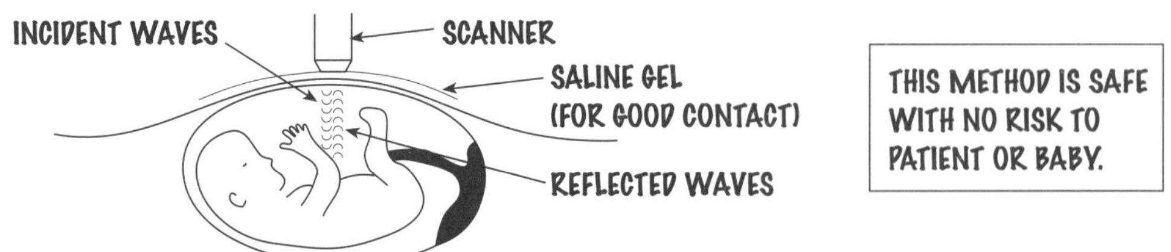

THIS METHOD IS SAFE WITH NO RISK TO PATIENT OR BABY.

2. ECHO - SOUNDING TO DETERMINE DEPTH OF WATER

- Ultrasonic waves are sent out from **BOTTOM** of **SHIP**.
- These waves are **REFLECTED BACK** from **SURFACE** at **BOTTOM** of **WATER** ...
- ... to a **DETECTOR**.
- The **TIME DELAY** of the **RELECTIONS** can be used to **CALCULATE** the **DEPTH** of the **WATER**.

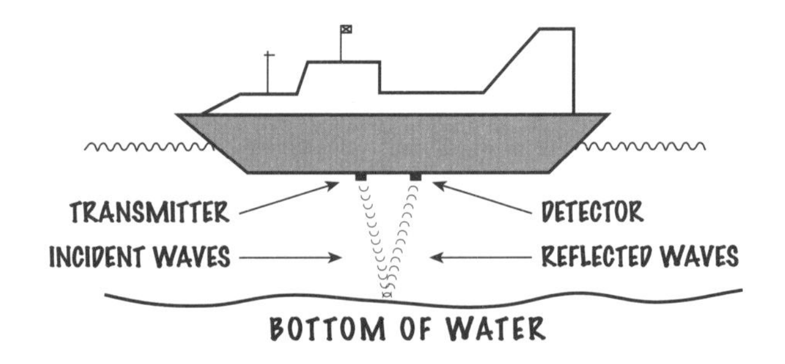

THIS METHOD CAN ALSO BE USED FOR LOCATING SUBMARINES AND SHOALS OF FISH.

3. ECHO - LOCATION IN BATS

- Ultrasonic waves are sent out by **BAT**.
- These waves are **REFLECTED BACK** from **ANY OBJECT IN THEIR PATH**.
- The reflected waves are **PROCESSED** by the bat ...
- ... resulting in the bat being able to '**VISUALISE**' **SIZE** and ...
- ... **DISTANCE TO** the reflecting object.

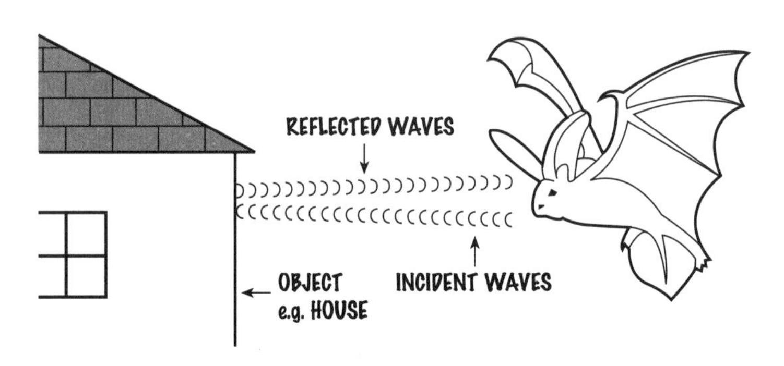

KEY POINTS:

- Ultrasound are sound waves of a frequency greater than 20,000Hz.

STRUCTURE OF THE EARTH

It is believed that the EARTH has a LAYERED STRUCTURE with ...

1. A THIN **CRUST** ...
... up to 10Km thick under oceans ...
... up to 100Km thick under continents.

2. A VISCOUS (SEMI-FLUID) **MANTLE** ...
... whose DENSITY INCREASES WITH DEPTH ...
... extending to almost halfway ...
... to the centre of the EARTH.

3. A **CORE** ...
... just over HALF THE EARTH'S DIAMETER ...
... with a LIQUID OUTER PART and ...
... a SOLID INNER PART.

CRUST
MANTLE
OUTER CORE
INNER CORE

EVIDENCE FOR THE STRUCTURE

Evidence for the LAYERED STRUCTURE has been gained by the study of EARTHQUAKES. These occur due to fracture of large masses of rock inside the Earth. The energy which is released TRAVELS through the earth as a series of shock waves called SEISMIC WAVES which are detected using SEISMOGRAPHS. There are TWO types of seismic waves.

P WAVES
• LONGITUDINAL • PASS THROUGH SOLIDS ... • ... AND LIQUIDS. • FASTER THAN S WAVES • SPEED INCREASES IN ... • ... MORE DENSE MATERIAL.

S WAVES
• TRANSVERSE. • PASS THROUGH SOLIDS ONLY. • SLOWER THAN P WAVES. • SPEED INCREASES IN ... • ... MORE DENSE MATERIAL.

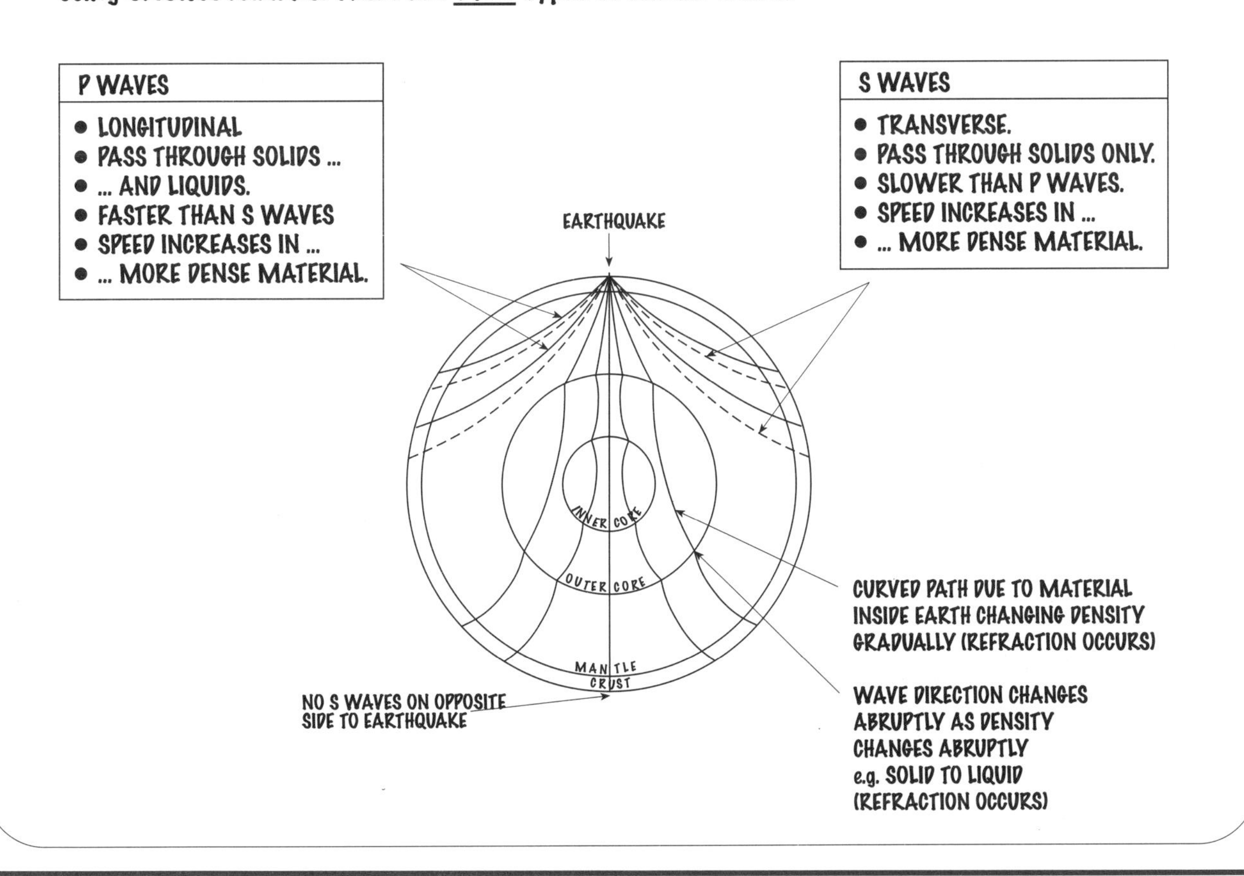

KEY POINTS:

• The Earth is made up of a Crust, Mantle and Core. • Earthquakes release shock waves called seismic waves. • There are two types: P Waves and S Waves.

WAVES SUMMARY QUESTIONS

1. What is a wave?
2. There are TWO types. Name them and give two examples of each.
3. What is a) amplitude b) wavelength c) frequency?
4. What is a) amplitude b) wavelength c) frequency measured in?
5. What is the wave equation?
6. Repeat example 2 (W.1) by using values for frequency from your newspaper to calculate the wavelength of the waves (velocity of radio waves is 300,000,000 m/s).
7. Draw a diagram to show the reflection of light.
8. What two things are always the SAME when light is reflected?
9. Draw a diagram to show the refraction of light.
10. Why does refraction occur?
11. Draw diagrams to show the diffraction of waves.
12. Under what conditions can light be diffracted?
13. What is total internal reflection?
14. What is critical angle?
15. What is an optical fibre?
16. How does an endoscope work?
17. Draw diagrams to show how a prism can reflect light through a) 90° b) 180°.
18. List the E-M spectrum in order of increasing frequency.
19. Name one use for each type of E-M radiation.
20. How are a) microwaves b) infra-red waves and c) ultra violet waves dangerous?
21. How is sound produced?
22. What do a) amplitude and b) frequency determine for sound?
23. What is a) a microphone b) a loudspeaker?
24. What are ultrasonic waves?
25. Name 3 uses for ultrasonic waves.
26. Describe the internal structure of the Earth?
27. What are earthquakes?
28. What are P waves?
29. What are S waves?
30. Draw a structural diagram of the earth showing the passage of P and S waves from an earthquake centre.

RADIOACTIVITY

CHARACTERISTICS AND DETECTION I — R 1

RADIOACTIVITY...

- ... is the SPONTANEOUS (i.e. naturally occurring) EMISSION of ENERGY from atomic nuclei ...
- ... as a result of the BREAKDOWN of UNSTABLE NUCLEI and it is ...
- ... a random process which is UNAFFECTED by any factors, PHYSICAL or CHEMICAL!

RADIATION can be detected by ...

1. GEIGER-MULLER TUBE

Argon gas, Wire, RADIATION, TO COUNTER, Tube, Pulse of current

- When RADIATION enters TUBE it creates IONS between WIRE and TUBE.
- EFFECT is just like a 'PULSE of current which is registered by a COUNTER ...
- ... and so it's possible to MEASURE the AMOUNT of radiation which enters.

(YOU DO NOT NEED TO KNOW HOW A G-M TUBE WORKS)

2. PHOTOGRAPHIC FILM

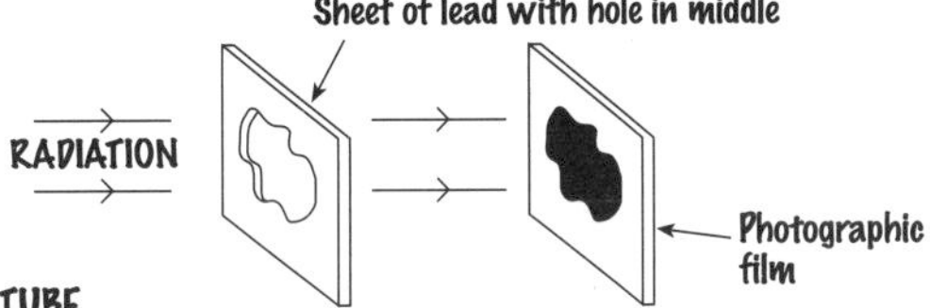

- PHOTOGRAPHIC FILM is 'BLACKENED' by RADIATION
- the MORE it is EXPOSED ...
- ... the 'BLACKER' the film.

BACKGROUND RADIATION

This is RADIATION THAT OCCURS ALL AROUND US.
The SOURCE of this radiation may be NATURAL or MAN-MADE (due to human activity).

NATURAL

1. ROCKS (e.g. granite) either below or above surface of Earth contain naturally occurring atoms which are radioactive. Decay may produce RADON GAS, also radioactive, which may seep into houses and be breathed in.
2. COSMIC RADIATION from outer space.

MAN MADE

1. MEDICAL, from the use of X rays mainly.
2. NUCLEAR INDUSTRY e.g. nuclear power stations including its waste material and fallout from weapon testing.

RADIOACTIVE EMISSIONS - types of and properties

There are THREE TYPES:	1. ALPHA (α) PARTICLES	2. BETA (β) PARTICLES	3. GAMMA (γ) RADIATION (part of the E-M spectrum - see W.4)
PENETRATING POWER:	LOW	MODERATE	HIGH
BLOCKED BY:	PAPER	THIN SHEETS OF ALUMINIUM	CONCRETE/LEAD

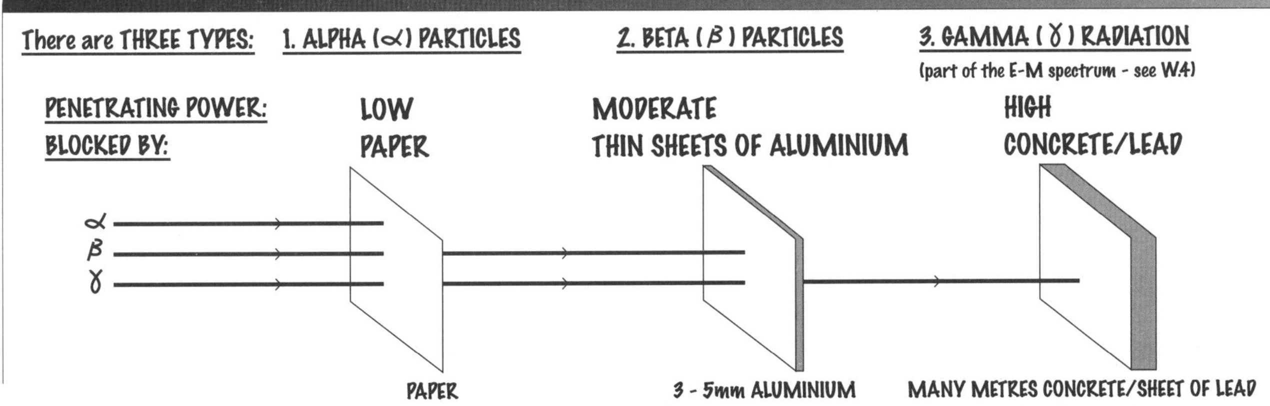

THEY ARE:	HELIUM NUCLEI (^4_2He) i.e. TWO PROTONS + 2 NEUTRONS	HIGH ENERGY ELECTRONS i.e. FAST MOVING	ELECTROMAGNETIC RADIATION WITH VERY HIGH FREQUENCY

RADIATION is EMITTED from UNSTABLE NUCLEI of atoms which are ISOTOPES of its STABLE ATOM where ...
... ISOTOPES are ATOMS which have the SAME No. OF PROTONS BUT DIFFERENT No. OF NEUTRONS as its STABLE ATOM.

KEY POINTS:

- Alpha particles, Beta particles and Gamma radiation are the three types of radioactive emissions.
- Radiation that occurs all around us is background radiation.

HALF-LIFE

This is the TIME it takes for:

where ○ = PARENT ATOM.
● = NEW ATOM FORMED AFTER PARENT ATOM HAS DECAYED.

- HALF A GIVEN NUMBER OF RADIOACTIVE ATOMS (PARENT ATOMS) ...
- ... TO DECAY TO DIFFERENT ATOMS.

e.g. (32) ○ HALF LIFE → (16) ○ + (16) ● HALF LIFE → (8) ○ + (24) ● HALF LIFE → (4) ○ + (28) ● ETC

However this definition of HALF-LIFE is very IMPRACTICAL and for calculations the following alternative definition is for more useful. The HALF-LIFE is the TIME taken for ...

- THE COUNT RATE (NUMBER OF PARENT ATOMS WHICH ARE DECAYING IN A CERTAIN TIME, MEASURED USING A G-M TUBE) ... TO FALL TO HALF ITS INITIAL VALUE.

EXAMPLE The table below shows the results of count rate against time for a radioactive isotope. From the in formation calculate the HALF-LIFE of the isotope.

TIME (minutes)	0	5	10	15	20	25	30
COUNT RATE (counts/minute)	200	160	124	100	80	62	50

There is a QUICK and a NOT SO QUICK way of calculating the HALF-LIFE ... however BOTH WAYS ARE EASY!!

The QUICK and EASY WAY

Remember ...
... HALF-LIFE = TIME FOR COUNT RATE TO HALVE.
From our table ...

COUNT RATE	TIME	HALF-LIFE
200 → 100	0 → 15	15 mins
160 → 80	5 → 20	15 mins
124 → 62	10 → 125	15 mins
	Average =	15 mins

The NOT SO QUICK but still EASY WAY!!

You will be asked to draw a graph of COUNT RATE against TIME.

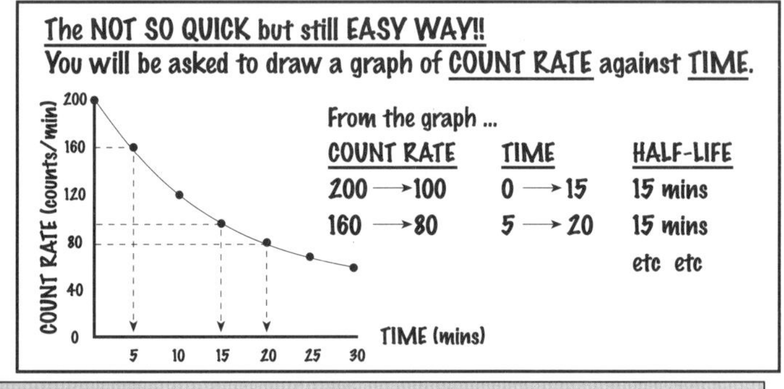

From the graph ...

COUNT RATE	TIME	HALF-LIFE
200 → 100	0 → 15	15 mins
160 → 80	5 → 20	15 mins
		etc etc

NB IF YOU ARE ALSO GIVEN THE BACKGROUND RADIATION COUNT RATE, YOU MUST TAKE THIS READING AWAY FROM EACH COUNT RATE TO GIVE THE CORRECTED COUNT RATE BEFORE YOU START TO WORK OUT THE HALF-LIFE.

DATING OF MATERIALS

- If the HALF-LIFE of an isotope is known then since certain materials contain RADIOISOTOPES which DECAY to produce STABLE ISOTOPES (they do not decay), then it is possible to DATE THE MATERIAL ...
- ... IGNEOUS ROCKS contain URANIUM -238 which decays.
- ... WOOD and BONES contain CARBON-14 which decays, when the organism dies!!

EXAMPLE

A very small sample of dead wood has a count rate of 1000 over a period of time, whereas the same mass of 'live' wood has a count rate of 4000 over an identical time period. If the half life of C-14 is 6000 years, calculate the age of the wood. WE NEED TO ASSUME THAT THE DEAD WOOD, WHEN ALIVE, WOULD ALSO HAVE HAD A COUNT RATE OF 4000 DUE TO C-14.

ORIGINAL COUNT RATE 4000 —HALF-LIFE→ 2000 —HALF-LIFE→ 1000 PRESENT COUNT RATE

Therefore the C-14 has taken 2 x HALF LIVES TO DECAY TO ITS CURRENT COUNT RATE.

Therefore age of wood = 2 x 6000
= 12000 years

KEY POINTS:

- The time taken for half a given number of radioactive atoms to decay to different atoms is called the half-life.

EFFECTS AND USES OF RADIATION

THE GOOD

1. MEDICINE
- To kill CANCEROUS CELLS ...
- ... use a γ source and a CALCULATED DOSE.
- STERILISING MEDICAL INSTRUMENTS ... kills all germs.
- As a TRACER inside body ...
- ... suitable SOURCE ...
- ... and HALF LIFE essential!!

2. INDUSTRY
- THICKNESS CONTROL for making PAPER and ALUMINIUM FOIL ...
- ... as amount of ABSORPTION depends on THICKNESS.
- As a TRACER to detect LEAKS in PIPES ...
- ... UPTAKE of RADIOACTIVE FERTILISERS by plants ...
- ... suitable SOURCE and HALF LIFE essential!!

THE BAD
- α, β and γ radiation can cause ...
- ... DAMAGE to ...
- ... LIVING CELLS in LIVING ORGANISMS ...

... AND THE UGLY!
- ... which can cause in ORGANS ...
- ... CANCER including LEUKAEMIA (cancer of the blood) and ...
- ... STERILITY or ABNORMALITIES IN CHILDREN BORN ...

An increase in the use of radioactive materials does produce ...
... SOCIAL ... ECONOMIC ... and ENVIRONMENTAL problems!

RADIOACTIVITY SUMMARY QUESTIONS

1. What is radioactivity?
2. Draw a Geiger-Muller tube.
3. How is photographic film affected by radiation?
4. What is background radiation?
5. Name 2 examples of a) NATURAL and b) MANMADE radiation sources.
6. What are the three types of radiation?
7. Draw a diagram to show how alpha, beta and gamma can be stopped.
8. What are a) alpha particles b) beta particles c) gamma rays?
9. What are isotopes?
10. What is half-life?
11. If a radioisotope has a half-life of 5 hours what fraction of the isotope will not have decayed after 10 hours?
12. If a radioisotope has a half-life of 2 hours what fraction of the isotope will have decayed after 10 hours?
13. The table below shows the results of count rate against time for a radioactive isotope.

Time (minutes)	0	10	20	30	40	50	60
Count rate (counts/minute)	400	330	250	200	164	122	96

Calculate the half-life using the QUICK method.

14. Draw a graph for the results above and use it to calculate the half-life.
15. If the BACKGROUND RADIATION count rate is 10 counts/minute, work out the corrected count rate for the results above and draw another graph to show corrected count rate against time. From your graph calculate the half-life.
16. How is it possible to date certain materials?
17. Repeat the bottom example on R.2 with a dead wood count rate of 1250 and a live one of 10,000.
18. Again repeat the same example with a dead wood count rate of 200 and a live one of 6,400.
19. What are the a) GOOD b) BAD and c) UGLY effects of radiation?
20. Name one example of how increased use of radioactive materials has produced a) social b) economic and c) environmental problems.

INDEX

PROGRESS AND REVISION CHART

PAGE No.	SECTION No.	CONTENT HEADING	SYLLAB. REF.	COVERED IN CLASS	REVISED	REVISED
5	FM1	Speed and Velocity.	(2.1)			
6	FM2	Force and acceleration I.	(2.2)			
7	FM3	Force and acceleration II.	(2.2)			
8	FM4	Force and non-uniform motion.	(2.3)			
9	FM5	Force on solids.	(2.4)			
10	FM6	Force and press. on solids and liquids.	(2.4)			
11	FM7	Force and press. on liquids and gases.	(2.4)			
12	FM8	The solar system I.	(4.1)			
13	FM9	The solar system II.	(4.1)			
14	FM10	The wider universe and the evolution of the universe and stars.	(4.2/4.3)			
16	E1	Energy resources I - non-renewables.	(5.1)			
17	E2	Energy resources II - renewables.	(5.1)			
18	E3	Energy resources III - the sun.	(5.1)			
19	E4	Energy transfers I - types and efficiency of.	(5.2)			
20	E5	Energy transfers II - conduction and convection.	(5.2)			
21	E6	Energy transfers III - radiation and evaporation.	(5.2)			
22	E7	Energy transfers IV - reducing heat losses.	(5.2)			
23	E8	Work, power and energy I.	(5.3)			
24	E9	Work, power and energy II - K.E. and P.E.	(5.3)			
26	EM1	Electric charge I.	(1.1)			
27	EM2	Electric charge II.	(1.1)			
28	EM3	Energy in circuits I.	(1.2)			
29	EM4	Energy in circuits II - series and parallel circuits.	(1.2)			
30	EM5	Energy in circuits III - resistance.	(1.2)			
31	EM6	Power and mains electricity I - fuses.	(1.3/1.7)			
32	EM7	Mains electricity II - plugs, insulation and earthing.	(1.7)			
33	EM8	Mains electricity III - paying for electricity.	(1.7)			
34	EM9	Electromagnetic forces I.	(1.4)			
35	EM10	Electromagnetic forces II.	(1.4)			
36	EM11	Electromagnetic induction.	(1.5)			
37	EM12	Transformers.	(1.6)			
39	W1	Characteristics of waves.	(3.1)			
40	W2	Light I.	(3.2)			
41	W3	Light II.	(3.2)			
42	W4	The electromagnetic spectrum.	(3.3)			
43	W5	Sound I.	(3.4)			
44	W6	Sound II - ultrasound.	(3.4)			
45	W7	Sound III - seismic waves.	(3.4)			
47	R1	Characteristics and detection I.	(6.1)			
48	R2	Characteristics and detection II.	(6.1)			
49	R3	Effects and uses of radiation.	(6.2)			